Springer Undergraduate Texts in Mathematics and Technology

Springer Undergraduate Texts in Mathematics and Technology (SUMAT) publishes textbooks aimed primarily at the undergraduate. Each text is designed principally for students who are considering careers either in the mathematical sciences or in technology-based areas such as engineering, finance, information technology and computer science, bioscience and medicine, optimization or industry. Texts aim to be accessible introductions to a wide range of core mathematical disciplines and their practical, real-world applications; and are fashioned both for course use and for independent study.

Nathan Grieve

An Introduction to Mathematical Programming and Network Science

Examples with Theory and Python

Springer

Nathan Grieve
School of Mathematics and Statistics
Carleton University
Ottawa, ON, Canada

ISSN 1867-5506 ISSN 1867-5514 (electronic)
Springer Undergraduate Texts in Mathematics and Technology
ISBN 978-3-032-13329-8 ISBN 978-3-032-13330-4 (eBook)
https://doi.org/10.1007/978-3-032-13330-4

Mathematics Subject Classification: 90-01, 90B06, 90B10, 90B50, 90C05, 90C08, 90C25, 90-04

This Springer imprint is published by the registered company Springer Nature Switzerland AG
The registered company address is: Gewerbestrasse 11, 6330 Cham, Switzerland

To Stacy.

Preface

What is Mathematical Programming and Network Science?

The real-world evolves rapidly—it is important to reorganize labour, manufacturing outputs, shipping distribution networks, supervisory and other managerial responsibilities accordingly. A particular challenge is to allocate resources in an effective and efficient way.

Achieving a scientific approach to studying and addressing such real-world challenges is one aim of *Mathematical Programming* and *Network Science*. The idea is to employ mathematical tools to achieve an optimal allocation of resources. For example, consider the case of a company that wishes to maximize its profits while at the same time minimizing its emissions outputs.

Some typical phases of a *modelling and programming work cycle* include

- defining a suitable *optimization problem* that is informed by relevant real-world problems and data;
- formulating a relevant mathematical *optimization model* to represent a given real-world problem;
- developing a computer-based procedure for deriving solutions to the mathematically formulated optimization problem;
- testing the mathematical model to see if the mathematically obtained solution is actually practical for the given real-world problem; and
- implementing conclusions and/or revising the model as required.

Mathematical programming and network science are very rich and difficult fields. They seek to solve mathematical optimization problems with the intent to apply such solutions to make informed real-world decisions for business, industry, government, non-profit and other organizational contexts. In doing so, they employ deep mathematical tools from linear algebra, combinatorial and convex analysis.

At the same time, there are ethical reflections that become relevant. For example, given knowledge of an optimal solution, is it socially responsible to act on it?

For instance, will maximizing profits inflict undue harm on a particular workforce? Similarly, will minimizing emissions outputs have undue harm on economic productivity?

Not withstanding such considerations, knowledge of and/or the ability to find, at least some optimal solutions to mathematical programming problems are of clear strategic interest to entities of all sizes and flavours. A detailed understanding, of even at least some subset, of the optimal and/or feasible solutions allows leaders and individuals to make informed socially responsible and scientifically guided decisions.

Aims, Intended Audience and How to Use This Text

The aim of this text is to provide a hands on practical introduction to concepts that are at the core of mathematical programming and network science. Here, we place emphasis on linear programming, the Simplex Method, and classical techniques from nonlinear convex mathematical programming.

The text seeks to reach upper year undergraduate students, beginning and midlevel graduate students, researchers and professionals who are studying and working within mathematics, technology, engineering, economics, business and related multidisciplinary quantitative fields. The text is suitable for a one-semester introductory undergraduate- and/or beginning graduate-level course. Portions of Chapters 1–9 comprise the core material for that purpose. At the same time, this text is suitable for self-study and as a reference resource.

In terms of prerequisites for this text, they are modest although some parts of the text become gradually more mathematically demanding than others. So, for example, it will be helpful if readers have taken a second course in linear algebra, but not necessarily a first course in abstract algebra. Similarly, it will be helpful if students have taken a course in multivariable calculus and an introductory course in probability and statistics, but they need not have taken a first course in real analysis nor an introductory course in topology nor a more advanced course in probability theory nor a more advanced course in higher dimensional statistics. Of course, as some parts of the text become more mathematically involved than others, some readers may wish to obtain familiarity with such introductory level advanced courses in tandem with their introduction to mathematical programming and network science.

In Appendix B, we summarize the most basic results about convexity and optimization of smooth functions in dimensions one and two. This helps to provide context and motivation for Chapters 9 and 10, which are the most mathematically engrossed in terms of real and convex analysis. In Appendix A, we provide a summary of the most basic and foundational results for solving systems of linear equations.

While our treatment of the Simplex Method is logically independent of Appendix A, readers may have an improved understanding and appreciation for the Simplex Method after having obtained a good grounding with solving systems of linear equations via the methods of Gaussian Elimination and/or Gauss-Jordan Elimination. Indeed, our approach to the Simplex Method, which is the key tool for solving linear programming problems, can be seen by analogy with Gaussian Elimination and/or Gauss-Jordan Elimination.

The core of the Simplex Method is documented in Chapter 5. This is mathematically more technical compared to other parts of the book. To illustrate our approach to the Simplex Method, in Chapter 6, see Section 6.1, we indicate how to implement this method directly in Python.

Navigating the Text, Logical Flow of Topics and Mathematical Sophistication

To assist with navigating the text, and as a guide for designing a semester course based on it, a suggested logical structure with chapter interdependence, logical flow of topics and mathematical sophistication is described in the following way.

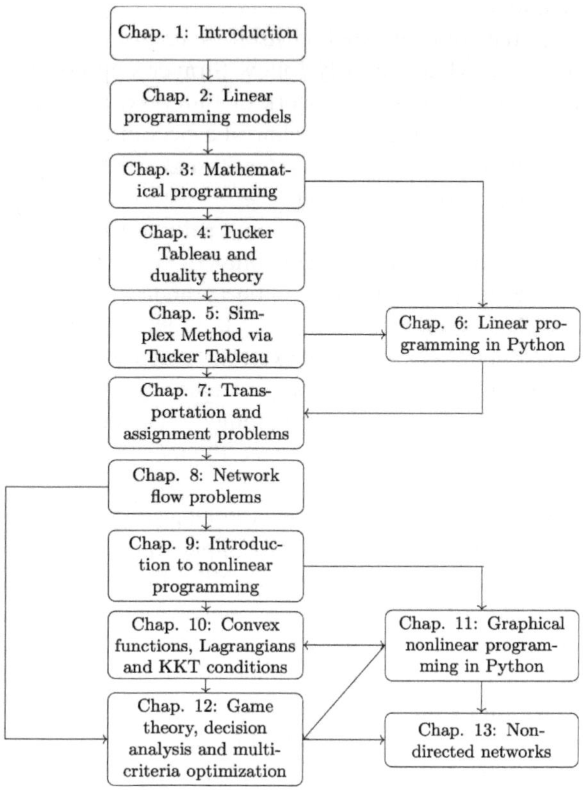

Viewpoint and Outline of the Text

My viewpoint builds on the texts of J. K. Strayer, *Linear programming and its applications*, V. Chvátal, *Linear programming*, F. S. Hillier and G. J. Lieberman, *Introduction to Operations Research*, R. T. Rockafellar, *Convex analysis*, W. L. Winston, *Operations Research: applications and algorithms*, R. Bronson and G. Naadimuthu, *Schaum's outline of Operations Research* (and others). It is with great pleasure that I acknowledge the influence that these texts have had on the present work. The informed reader who is familiar with these works will have no trouble in seeing their influence throughout my exposition here.

One of my goals is to make the many illustrative and representative examples, which arise in these influential works, timeless. So, for example, some of the many case study type examples treated in these works are presented here quite generally as models with abstract parameters. We also place emphasis on the wide applications of the Simplex Method and general purpose quadratic and separable convex nonlinear programming techniques.

My approach to the Simplex Method itself is very direct. This is exposed in Chapter 5. In doing so, while I mostly follow Strayer's approach, I also build on the work of Dantzig in addition to the work of Balinkski and Tucker. This part of Chapter 5, see Section 5.3, is the most technical aspect of the book. In Section 5.6, we place emphasis on *linear programming duality* and its interplay with the Simplex Method and the theory of Tucker Tableau.

First time readers may very well wish to omit entirely the technical details of the Simplex Method. Instead, they may prefer to immediately turn to the linear programming features of the Python package scipy.optimize and/or other linear programming solving platforms. This is a perfectly fine way to proceed. All of the mathematical programs that appear in the examples and exercises that are contained in this text can be solved using the features of the Python packages numpy, mathplotlib and scipy. We also illustrate how to solve such programming problems using the most basic features of the Algebraic Modelling Language Pyomo. (See Chapters 6 and 11 for further details.)

Readers with background and interest in computer programming are encouraged to give their own direct implementation of the Simplex Method in computer programming languages such as Python, C++ or Java. With this in mind, our presentation here of the Simplex Method, via the theory of Tucker Tableau, allows for an efficient and direct implementation in any of these programming languages. This is illustrated in Chapter 6 which includes a pragmatic implementation of Tucker Tableau and the Simplex Method within Python.

The Simplex Method, which is the key tool for solving linear programming problems, is analogous to the methods of Gaussian Elimination and Gauss-Jordan Elimination (which should be familiar from an introductory course in linear algebra). For completeness, we expose the most basic theory that surrounds those methods in Appendix A. However, it is important to note that our approach to the Simplex

Method, based on judicious pivots applied to Tucker Tableau, is entirely separate from Gaussian Elimination and Gauss-Jordan Elimination.

Another feature of the present text is to place emphasis on representative case study type examples. Oftentimes throughout the text, a general representative example is worked out; special instances of such models are discussed in the exercises and/or prior to formulating the general model in a more abstract way. It is hoped that this pedagogical approach gives the reader an intuition and understanding of a general purpose structured approach to studying and understanding the key representative models that drive the theory.

The text is mostly logically self-contained and has only modest required background. Having said this, especially first time readers, may wish to omit reading of the proofs entirely. Indeed, while we do state and prove the most significant of theorems and algorithms which form the core foundations for mathematical programming and network science, our emphasis here is to understand how these powerful theoretical tools can be applied to concrete real-world problems. For instance, while we do explain the intuition and key ideas as to why these theoretical tools indeed hold true, we also provide references to the mathematical literature for additional context and details. Given the hands on introductory approach that we are pursuing here, the interested mathematically mature and well-motivated student will have no trouble following the key ideas that are given in these primary and expository sources.

What's Next?

There are several multidisciplinary directions for readers to explore having mastered the material that is presented here. For this purpose, some complementary texts include J. Brinkhuis *Convex Analysis for Optimization*, M. Ehrgott *Multicriteria optimization*, H. A. Eiselt and C.-L. Sandblom *Linear programming and its applications*, B. Sagan *Combinatorics: the art of counting*, M. L. Bynum, G. A. Hackebeil, W. E. Hart, C. D. Laird, B. L. Nicholson, J. D. Siirola, J.-P. Watson and D. L. Woodruff *Pyomo—optimization modeling in Python*, G. M. Ziegler *Lectures on polytopes* and many others. These texts have shaped the exposition here too.

Acknowledgements

It is my pleasure to thank my teachers, colleagues, students and the many authors from whom I have learned a great deal from. While preparing this work, I was supported by a Discovery Grant from the Natural Sciences and Engineering Research Council of Canada. I especially thank Elizabeth Loew from Springer Nature for her enthusiastic interest and facilitation regarding this work. Finally, I thank anonymous reviewers for their insightful suggestions.

Orléans, ON, Canada
October 2025

Nathan Grieve

Competing Interests The author has no competing interests to declare that are relevant to the content of this manuscript.

Contents

Chapter 1
Introduction

Overview

- Guiding examples.
- Introduction to mathematical programming and mathematical optimization.
- Graphical method for solving two variable linear programming problems.
- Additional examples of mathematical programs.

1.1 Guiding Examples

We begin with a collection of real-world examples which naturally give rise to *mathematical programming problems*.

Example 1.1. An airline traveller needs to decide among a collection of items to pack in their checked baggage subject to the airline's weight limit. Which items should be taken in order to maximize the traveller's total preference value without exceeding the airline's imposed weight restriction?

Example 1.2. A banquet catering company needs to decide the amount of a given collection of fruit juices to include in the punch. In order to ensure a good quality of the punch, there are constraints on the amount of each particular fruit juice that can be used within the punch. What allocation of each juice will maximize the punch quality? What allocation will minimize the cost?

Example 1.3. A company manufactures two types of products. Each requires a certain number of hours of work in the parts division and a certain number of hours of work in the assembly division. On the other hand, the parts and assembly division are, respectively, open only for a certain number of hours in each particular day. How many of each type of product should be manufactured in order to maximize the expected profit?

N. Grieve, *An Introduction to Mathematical Programming and Network Science*,
Springer Undergraduate Texts in Mathematics and Technology,
https://doi.org/10.1007/978-3-032-13330-4_1

Example 1.4. A manufacturing company has two factories each of which is able to produce two types of products. Both factories have production output constraints whereas each of the two products have consumer product demand constraints. What is the optimal number of days that each factory should be operated in order for the company to minimize its production costs?

Example 1.5. A company has a collection of factories that supply a collection of warehouses via a central distribution centre. Each factory has production constraints, each warehouse has consumer demand constraints and finally the shipping distribution network has capacity constraints. Which choice of shipping route will minimize the shipping cost while at the same time meet the demand for each of its products?

Example 1.6. An investor is choosing between a collection of financial products to invest in. How should the investor's capital be allocated among these products in order to maximize the expected rate of return and minimize the expected amount of risk?

Example 1.7. A delivery truck must make a collection deliveries, visiting each location only once, before returning to the warehouse. What ordering of the deliveries will minimize the total distance that the delivery truck must travel?

Example 1.8. A collection of print jobs must be printed in a single print shop. Each print job has a deadline. What order should the print shop print the orders in order to ensure that each job is completed prior to its deadline?

Example 1.9. A fibre artist must allocate different types of yarn for a knitting project. Which allocation will meet the required size, colour, durability and blend requirements while at the same time achieve a minimal cost?

1.2 Introduction to Mathematical Programming

The fields of *Mathematical Programming* and *network science* provide mathematical frameworks for solving real-world *optimization problems* such as the examples that are described in Section 1.1. A key role is played by *mathematical optimization programming problems*.

They seek to *optimize*, that is *maximize* and/or *minimize*, a given real-valued *objective function*

$$Z = F(x_1, \ldots, x_n)$$

subject to a given collection of *constraints*

$$G_i(x_1, \ldots, x_n) \leq, = \text{ or } \geq b_i \text{ for } i = 1, \ldots, N.$$

The unknown variables $x_1, \ldots, x_n$ are referred to as *decision variables*. There may be additional restrictions. For instance, some or all of the decision variables may be required to be nonnegative integers.

The idea is that the optimal solution(s) of such mathematical programming problems—should any exist—yield optimal solution(s) to the given real-world optimization problem. To get a sense, in Sections 1.3, 1.5 and 1.6, we discuss a collection of highly representative mathematical optimization programming problems. In Section 1.4, we describe a direct graphical approach for solving two variable linear programming problems.

In the meantime, here are two concrete examples of mathematical programming problems.

Example 1.10. • Maximize the linear objective function

$$f(x_1, x_2) = 3x_1 + 2x_2$$

• Subject to the constraints

$$3x_1 + 2x_2 \leq 6$$

and

$$x_1, x_2 \geq 0.$$

Example 1.11. • Maximize the quadratic objective function

$$f(x_1, x_2) = 4x_1 - x_1^2 + 3x_2 - 2x_2^2$$

• Subject to the constraints

$$3x_1 + 2x_2 \leq 6$$

and

$$x_1, x_2 \geq 0.$$

Example 1.10 is an instance of a *linear maximization linear programming problem* whereas Example 1.11 is a *quadratic maximization linear programming problem.* Indeed, Example 1.10 has a linear objective function and linear constraints; Example 1.11 has a quadratic objective function and linear constraints.

1.3 Linear Mathematical Programming Problems

An extremely important class of mathematical programming problems occurs when the objection function

$$Z = F(x_1, \ldots, x_n)$$

together with all of the constraint functions

$$G_i(x_1, \ldots, x_n) \text{ for } i = 1, \ldots, N$$

are *linear*. We discuss a collection of illustrative examples of linear programming problems in what follows. Some representative general purpose models are described in Section 1.5 and Chapter 2.

Example 1.12. A contractor builds decks and installs patio stones. Before starting a job, the contractor must make arrangements to acquire the materials from the hardware store. The contractor is able to visit the hardware store during at most 10 hours in a week. It typically takes 1 hour of work to acquire deck building materials and 1.5 hours of work to acquire patio stone materials.

Suppose that the contractor is able to build decks and install patio stones for at most 35 hours a week. Further, suppose that a typical deck build requires 15 hours of work whereas a typical patio stone installation requires 17 hours of work. Finally, suppose that building a deck yields \$3, 500 of profit and that installing a patio yields \$4, 500 of profit. In order to maximize profits, how many decks should the contractor build and how many patios should be installed?

An optimal solution can be found via a suitably formulated maximization linear programming problem. To this end, let x_1 be the number of decks to build and let x_2 be the number of patios to install. The contractor is able to visit the hardware store for at most 10 hours in a week and can do construction and/or installation for at most 35 hours per week.

The objective function to maximize is the profit function

$$P(x_1, x_2) = 3500x_1 + 4500x_2.$$

It typically takes 1 hour of work to acquire the deck building materials and 1.5 hours of work to acquire the patio stone materials. It takes 15 hours of work to build a deck and 17 hours of work to install patio stones.

The programming problem's constraints are thus

$$\begin{aligned} x_1 \ + 1.5x_2 &\le 10 \\ 15x_1 + \ 17x_2 &\le 35 \end{aligned}$$

and

$$x_1, x_2 \ge 0.$$

Upon solving this programming problem, for instance, using the techniques that are described in Section 1.4, it turns out that the contractor should install about 2 patios a week which will yield a profit of about \$9, 265.

Example 1.13. A hardware store has two local suppliers. Each supplies deck building and patio stone installation materials. One supplier is able to deliver materials for 20 decks per week and 15 patios. The other is able to deliver materials for 18 decks and 19 patios.

In order to meet production demands throughout the spring/summer building season, the hardware store requires materials for at least 150 decks and at least 180 patios. Finally, suppose that it costs \$2, 500 per weekly delivery from supplier one and \$2, 900 per weekly delivery from supplier two. How many deliveries of each respective supplier should the hardware store request in order to minimize the delivery cost?

An optimal solution can be found via a suitable minimization linear programming problem. To this end, let y_1 be the number of deliveries from supplier one and let y_2 be the number of deliveries from supplier two.

The objective function to minimize is the cost function

$$C(y_1, y_2) = 2,500y_1 + 2,900y_2.$$

On the other hand, the hardware store needs inventory for at least 150 decks and at least 180 patios. The first supplier is able to provide materials for 20 decks per week whereas the second supplier is able to provide materials for 18 decks per week. Similarly, the first supplier is able to provide materials for 15 patios and the second supplier is able to provide materials for 19.

The programming problem's constraints are thus

$$\begin{aligned} 20y_1 + 18y_2 &\geq 150 \\ 15y_1 + 19y_2 &\geq 180 \end{aligned}$$

and

$$y_1, y_2 \geq 0.$$

Upon solving this programming problem, for instance, using the techniques that are described in Section 1.4, it turns out that the hardware store owner should aim to have about 9.5 deliveries from supplier two and no deliveries from supplier one. The projected total cost for the shipments is \$27, 473.

Examples 1.14 and 1.15 below provide two representative examples of optimization problems that can be solved via linear mathematical programming problems. They complement and are more general formulations of Examples 1.3 and 1.4 from Section 1.1. They, respectively, generalize Examples 1.12 and 1.13.

Example 1.14 (Maximize profits given production allocation constraints)**.** A company manufactures two types of products. Each product requires a different amount of work in the parts and assembly divisions. Respectively, label these two products as Type 1 and Type 2. Let us refer to the parts division as Division 1 and the assembly division as Division 2.

Further, suppose that

- the production of a single Type 1 product requires a_{11} hours of work in the parts division (Division 1) and a_{21} hours of work in the assembly division (Division 2);

- the production of a single Type 2 product requires a_{12} hours of work in the parts division (Division 1) and a_{22} hours of work in the assembly division (Division 2);
- the parts division (Division 1) is operated for at most b_1 hours per day;
- the assembly division (Division 2) is operated for at most b_2 hours per day;
- the profit upon sale per Type 1 product is c_1 dollars; and
- the profit upon sale per Type 2 product is c_2 dollars.

Within this context, the underlying optimization problem is to determine the number of Type 1 and Type 2 products that must be manufactured per day in order to maximize profits.

To this end, let x_1 denote the number of Type 1 products manufactured per day. Let x_2 be the number of Type 2 products manufactured per day. In our mathematical model, we allow the possibility of noninteger values for x_1 and x_2. Practically speaking, this means that it is possible to produce partial amounts of Type 1 and Type 2 products on any given day.

We aim to maximize profit. In mathematical terms, this means that we want to maximize the *profit function*

$$P(x_1, x_2) = c_1x_1 + c_2x_2$$

subject to the constraints that

- the parts division is available for at most b_1 hours per day;
- the assembly division is available for at most b_2 hours per day; and
- it is not possible to produce a negative number of Type 1 or Type 2 products in a given day.

These constraints are expressed mathematically as

$$\begin{aligned} a_{11}x_1 + a_{12}x_2 &\leq b_1 \\ a_{21}x_1 + a_{22}x_2 &\leq b_2 \end{aligned}$$

and

$$x_1, x_2 \geq 0.$$

Here, the first constraint arises from the fact that the Type 1 product requires a_{11} hours of time in the parts division, whereas the Type 2 product requires only a_{12}. Similarly, the second constraint stems from the fact that the Type 1 product requires a_{21} hours in the assembly division, whereas the Type 2 product requires a_{22}.

Finally, the last two constraints, which are oftentimes referred to as the *implied constraints*, arise from the practical consideration that it is not possible to produce a negative number of products in any given day. An additional requirement could be that x_1 and x_2 are integers. What this would represent is that it is not possible to produce noninteger amounts of product in any given day.

Putting everything together, a solution to the given profit maximization problem may be expressed as the following *maximization linear programming problem*.

- Maximize the linear objective function

$$P(x_1, x_2) = c_1x_1 + c_2x_2$$

- Subject to the constraints

$$\begin{aligned} a_{11}x_1 + a_{12}x_2 &\leq b_1 \\ a_{21}x_1 + a_{22}x_2 &\leq b_2 \end{aligned}$$

and

$$x_1, x_2 \geq 0.$$

In Example 1.17, we explain a direct graphical approach for solving such 2×2 maximization programming problems.

In Example 1.15 below, instead of maximizing profits given certain allocation production constraints, as in Example 1.14, we seek to minimize costs given certain production constraints.

Example 1.15 (Minimize costs given production constraints)**.** A company has two factories, Factory 1 and Factory 2, each of which is able to produce two types of products, Type 1 and Type 2.

Suppose further that

- Factory 1 is able to produce a_{11} Type 1 products per day and a_{12} Type 2 products per day, whereas Factory 2 is able to produce a_{21} Type 1 per day and a_{22} Type 2 products per day;
- in order to meet production demands, the company requires at least c_1 Type 1 products and at least c_2 Type 2 products; and
- it costs s_1 dollars per day to operate Factory 1 and s_2 dollars per day to operate Factory 2.

The underlying optimization problem is to determine the optimal number of days that each factory should be operated in order for the company to minimize its production costs.

Let y_1 be the number of days that Factory 1 is open and let y_2 be the number of days that Factory 2 is open. The aim is then to minimize the *cost function*

$$C(y_1, y_2) = s_1y_1 + s_2y_2$$

subject to the constraints that

- a total of c_1 Type 1 products are required;
- a total of c_2 Type 2 products are required; and
- it is not possible for the factories to be open for a negative number of hours in a given day, whereas the factories are allowed to be open for part days.

In mathematical terms, these constraints can be expressed as

$$a_{11}y_1 + a_{21}y_2 \geq c_1$$
$$a_{12}y_1 + a_{22}y_2 \geq c_2$$

and

$$y_1, y_2 \geq 0.$$

The first constraint arises from the fact that Factory 1 is able to produce a_{11} Type 1 products, Factory 2 is able to produce a_{21} Type 1 products and finally c_1 Type 1 products are required. Similarly, the second constraint arises from the fact that Factory 1 is able to produce a_{12} Type 2 products, Factory 2 is able to produce a_{22} Type 2 products and c_2 Type 2 products are required.

Finally, the last two nonnegativity constraints arise from the practical considerations that the factories cannot be open for a negative number of days. On the other hand, since the factories are allowed to be open for part days, there is no requirement that y_1 and y_2 must be integers.

Putting everything together, a solution to the given cost minimization problem may be expressed as the following *minimization linear programming problem*.

- Minimize the linear objective function

$$C(y_1, y_2) = s_1y_1 + s_2y_2$$

- Subject to the constraints

$$a_{11}y_1 + a_{21}y_2 \geq c_1$$
$$a_{12}y_1 + a_{22}y_2 \geq c_2$$

  and

$$y_1, y_2 \geq 0.$$

In Section 1.4, we explain a direct graphical approach for solving two variable linear programming problems. Before proceeding in that direction, in Example 1.16 below, we describe a particular instance of the so-called *knapsack problem*. This is achieved via an *air traveller's packing dilemma*. (Compare, for instance, with Example 1.1.)

Example 1.16 (Air traveller's dilemma)**.** Consider the case that an airline traveller wants to take a collection of at most n items subject to an airline's b lb weight limit. To aid with the decision process, the traveller assigns a value of importance to each of the n items. The relevant data is summarized in Table 1.1.

Which items should be taken in order to maximize the total preference value without exceeding the airline's imposed weight restriction?

Let x_i, for $i = 1, \ldots, n$, denote the amount of item i to be taken and consider the objective function

Table 1.1 Data table for Example 1.16

Item	1	...	n
Weight (lb)	w_1	...	w_n
Value	v_1	...	v_n

$$f(x_1, \ldots, x_n) = v_1 x_1 + \cdots + v_n x_n.$$

The weight restriction is

$$w_1 x_1 + \cdots + w_n x_n \le b$$

and thus a solution to the traveller's dilemma is given as a solution to the following maximization linear mathematical program.

- Maximize the objective function

$$f(x_1, \ldots, x_n) = v_1 x_1 + \cdots + v_n x_n$$

- Subject to the constraints

$$w_1 x_1 + \cdots + w_n x_n \le b$$

  and

$$x_i = 0 \text{ or } 1 \text{ for } i = 1, \ldots, n.$$

One feature of this example is the requirement that the decision variables x_i must be either 0 or 1. This is a general feature of *scheduling and assignment type problems*. (See Section 7.4.)

1.4 Graphical Method for Solving Two Variable Linear Programming Problems

In this section, we describe a direct *graphical approach* for solving two variable linear programming problems. (See Examples 1.17 and 1.18.) The approach that is illustrated within these examples generalizes to give a practical method for solving $m \times 2$ linear programming problems.

It turns out that the theoretical justification for this approach is deduced as a special case of the more general and powerful geometric method for solving a large class of linear programming problems. We refer to Sections 3.6 and 3.7 for more details. As is explained in Chapter 5, these theoretical considerations ultimately underlie the key ideas that make the Simplex Method possible.

Table 1.2 The extreme points and objective function values for Example 1.17

Extreme point	Value of $P(x, y)$ at extreme point
$(0, 0)$	0
$(5, 0)$	1250
$(0, 5)$	2000
$(3, 4)$	2350

Example 1.17. Consider the following 2×2 maximization linear programming problem.

- Maximize the linear objective function

$$P(x, y) = 250x + 400y$$

- Subject to the constraints

$$\begin{aligned} 2x + \ y &\leq 10 \\ x + 3y &\leq 15 \end{aligned}$$

  and

$$x, y \geq 0.$$

Note that this a special instance of the general 2×2 maximization profits linear program model that is formulated in Example 1.14.

By examining these above constraint inequalities in the plane, the *extreme points* of the *constraint set* (equivalently, the *corner points* of the *feasible region*) are easily seen to be

$$(0, 0), (5, 0), (0, 5) \text{ and } (3, 4) \in \mathbb{R}^2.$$

The given linear programming problem can be solved by evaluating the objective function $P(x, y)$ at each of these extreme points and then noting which evaluations yield maximums. This is illustrated in Table 1.2.

Thus, $P(x, y)$ is maximized at the extreme point $(3, 4) \in \mathbb{R}^2$. The feasible region is illustrated in Figure 1.1.

Example 1.18. Consider the following 2×2 minimization linear programming program.

- Minimize the linear objective function

$$C(x, y) = 300x + 500y$$

- Subject to the constraints

$$\begin{aligned} 2x + 4y &\geq 100 \\ 5x + 4y &\geq 160 \end{aligned}$$

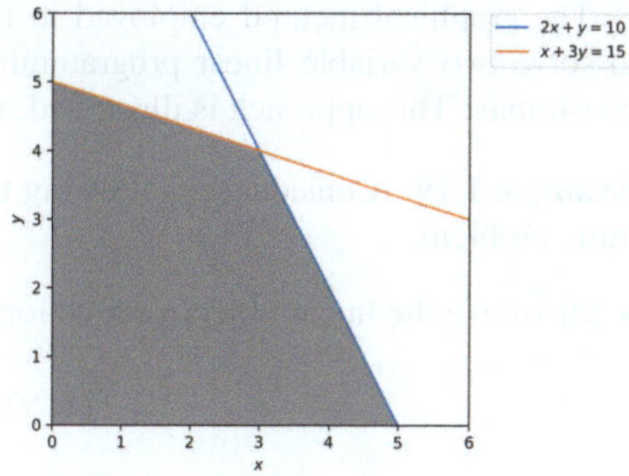

Fig. 1.1 The feasible region for Example 1.17

and

$$x,y \geq 0.$$

Here, the feasible region is *unbounded*. This is illustrated in Figure 1.2.

However, the objective function $C(x, y)$ is bounded from below. Further, the extreme points of the feasible region are

$$(50, 0), (0, 40) \text{ and } (20, 15) \in \mathbb{R}^2.$$

Hence a solution to the given minimization linear programming problem can be found by evaluating $C(x, y)$ at each extreme point and then observing which evaluation(s) yield minimum(s). This is illustrated in Table 1.3.

Therefore, $C(x, y)$ is minimized at the extreme point $(20, 15) \in \mathbb{R}^2$.

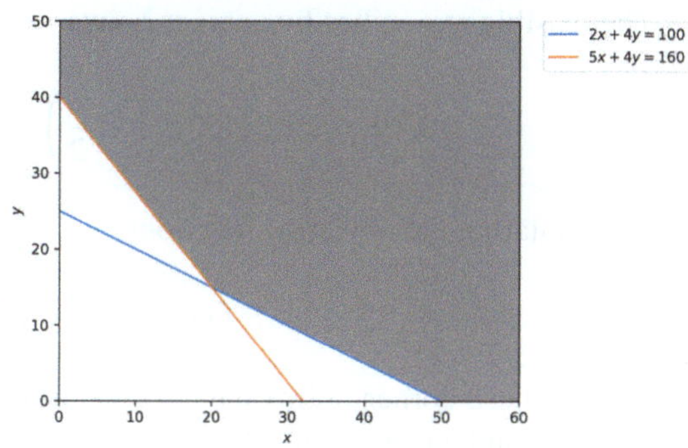

Fig. 1.2 The feasible region for Example 1.18

Table 1.3 The extreme points and objective function values for Example 1.18

Extreme point	Value of $C(x, y)$ at extreme point
(50, 0)	15000
(0, 40)	20000
(20, 15)	13500

The graphical method employed in Examples 1.17 and 1.18 can also be used to solve two variable linear programming problems with an arbitrary number of constraints. This approach is illustrated in Example 1.19.

Example 1.19. Consider the following two variable maximization linear programming problem.

- Maximize the linear objective function

$$f(x, y) = 4x + 7y$$

- Subject to the constraints

$$\begin{aligned} x \quad\quad &\le 4 \\ y &\le 5 \\ 3x + 2y &\le 15 \end{aligned}$$

and

$$x, y \ge 0.$$

As in Examples 1.17 and 1.18, the idea is to consider the value of the objective function at each of the corner points of the feasible region. The key point again is that the objective function is maximized at some corner point of the feasible region. This is implied by general theory, but is also something that can be checked directly on this particular example.

Now, in order to determine all corner points of the feasible region, the idea is to study the simultaneous intersection of each of the *half spaces* that are defined by each of the programming problem's linear constraints. This region in $\mathbb{R}^2$ is the programming problem's *feasible region.*

The feasible region has five corner points

$$(0, 0), (0, 5), (4, 0), \left(4, \frac{3}{2}\right) \text{ and } \left(\frac{5}{3}, 5\right) \in \mathbb{R}^2.$$

By evaluating the objective function

$$f(x, y) = 4x + 7y$$

at each of these corner points the conclusion is that it is maximized at the corner point $\left(\frac{5}{3}, 5\right)$ and has optimal value

$$f\left(\frac{5}{3}, 5\right) = \frac{125}{3} \approx 41.67.$$

The feasible region is illustrated in Figure 1.3.

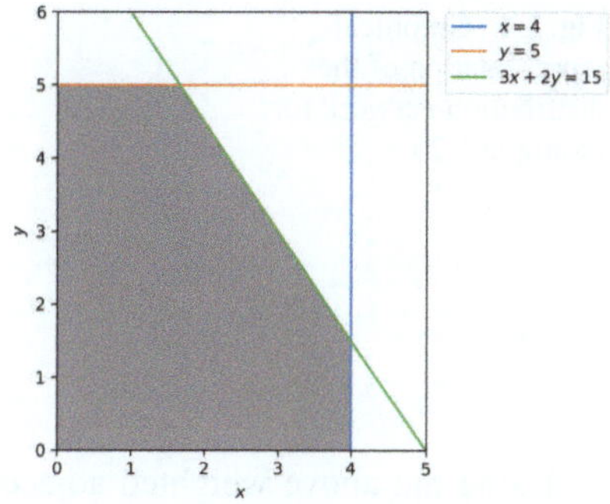

Fig. 1.3 The feasible region for Example 1.19

1.5 Network Flow Problems

Another rich source of linear programming problems arises from *network flow problems*. These are discussed in detail in Chapter 8. The main considerations are illustrated in Example 1.20 below. (Compare with Example 1.5.) A more general model for network flow problems is described in Example 1.21.

Example 1.20. Suppose that a company has two factories which supply two warehouses via a central distribution centre. Respectively, let v_1 denote Factory 1, v_2 Factory 2, v_3 the distribution centre, v_4 the first warehouse and v_5 the second warehouse.

The network is set up in such a way that Factory 1 is able to supply products to the first warehouse directly, all of the products produced from the Factory 2 are sent to the distribution centre and some products are able to flow from Factory 1 to Factory 2. Further, some products are able to flow between the two warehouses. Finally, the distribution centre supplies products to the first warehouse only through the second warehouse.

We say that $v_1, \ldots, v_5$ are the *nodes* of the network. The supply cost per unit of the distribution network can be represented as a *weighted digraph* with *weight adjacency table* given as in Table 1.4.

Table 1.4 Data table for Example 1.20

	v_1	v_2	v_3	v_4	v_5
v_1	0	2	4	9	0
v_2	0	0	3	0	0
v_3	0	0	0	0	1
v_4	0	0	0	0	3
v_5	0	0	0	2	0

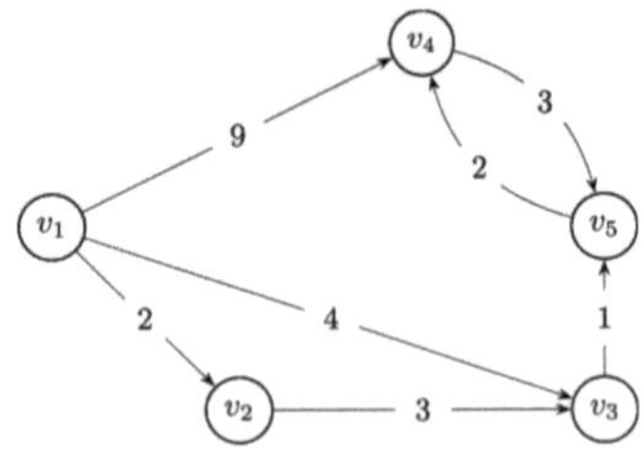

Fig. 1.4 Graphical representation of the distribution network for Example 1.20

Using the above weighted adjacency matrix, it follows that node v_i can supply units to node v_j at a cost of v_{ij} hundreds of dollars per unit. The network with labelled edges is depicted in Figure 1.4.

In terms of inventory, Factory 1 has produced 50 units, Factory 2 has produced 40 units and Warehouse 1 needs 30 units whereas Warehouse 2 needs 60 units. The network has five net flow constraints one for each of the five network nodes.

The key underlying network flow assumption is that the amount shipped out minus the amount shipped in must equal the required amount. Inherent in the given distribution network are two capacity constraints that at most 10 units can be shipped from Factory 1 to Factory 2 and at most 80 units can be shipped from the distribution centre to Warehouse 2.

In order to formulate a suitable linear programming problem, consider seven decision variables x_{12}, x_{13}, x_{14}, x_{23}, x_{35}, x_{45} and x_{54}. It turns out that each variable corresponds to an edge in the weighted digraph that models the distribution network.

The aim is to minimize the objective function

$$Z = 2x_{12} + 4x_{13} + 9x_{14} + 3x_{23} + x_{35} + 3x_{45} + 2x_{54}$$

subject to the problem's constraints. There are three kinds of constraints. These are the network flow constraints, the upper bound capacity constraints and the nonnegativity constraints. They are formulated in the following way.

- **Network flow constraints** Denoting inventory by positive units and required inventory by negative units, the network flow constraints can be summarized succinctly in the following tabular form. (See Table 1.5.)

Table 1.5 Network flow constraints for Example 1.20

Network node	Constraint	Explanation
v_1	$x_{12} + x_{13} + x_{14} = 50$	Factory 1
v_2	$-x_{12} + x_{23} = 40$	Factory 2
v_3	$-x_{13} - x_{23} + x_{35} = 0$	Distribution Centre
v_4	$-x_{14} + x_{45} - x_{53} = -30$	Warehouse 1
v_5	$-x_{35} - x_{45} + x_{54} = -60$	Warehouse 2

- **Upper bound capacity constraints**

$$x_{12} \leq 10 \text{ (Capacity flow constraint from Factory 1 to Factory 2)}$$

$$x_{35} \leq 80 \text{ (Capacity flow constraint from distribution centre to Warehouse 2).}$$

- **Nonnegativity constraints**

$$x_{12}, x_{13}, x_{14}, x_{23}, x_{35}, x_{45}, x_{54} \geq 0.$$

It turns out that an optimal solution to this linear program can be found as an application of the Simplex Method. The conclusion is that an optimal solution occurs when

$$x_{12} = 0, x_{13} = 40, x_{14} = 10, x_{23} = 40, x_{35} = 80, x_{45} = 0 \text{ and } x_{54} = 20;$$

the resulting optimal shipping cost is then

$$Z = \$49,000.$$

A more general model for distribution of goods through a distribution network is given in Example 1.21 below.

Example 1.21 (Distribution of goods through a distribution network)**.** A company has n nodes $v_1, \ldots, v_n$ in a distribution network. Suppose that the supply cost per unit of the distribution is described via the *weight adjacency table* which is given in Table 1.6.

Using the above weighted adjacency matrix, what is meant is that node v_i can supply units to node v_j at a cost of v_{ij} units of currency per unit. Here, v_{ij} is the entry of the above table that is indexed by row i and column j.

It turns out that in the diagrammatic description of the distribution network, there is a *directed edge* from vertex v_i to vertex v_j whenever the table has a nonzero entry v_{ij} in the ith row and the jth column. By convention $v_{ii} = 0$ for all $i = 1, \ldots, n$, since no node v_i is able to supply itself with the particular product. Graphically, the network has no directed loops.

The other table that is of interest, see Table 1.7, governs the *network flow capacity*—that is, the amount of units c_{ij} that can flow from node v_i to v_j.

Table 1.6 Weight adjacency data table for Example 1.21

	v_1	$\ldots$	v_n
v_1	v_{11}	$\ldots$	v_{1n}
$\vdots$	$\vdots$	$\ddots$	$\vdots$
v_n	v_{n1}	$\ldots$	v_{nn}

Table 1.7 Network flow capacity data table for Example 1.21

	v_1	$\dots$	v_n
v_1	c_{11}	$\dots$	c_{1n}
$\vdots$	$\vdots$	$\ddots$	$\vdots$
v_n	c_{n1}	$\dots$	c_{nn}

Table 1.8 Weight adjacency flow capacity data table for Example 1.21

	v_1	$\dots$	v_n
v_1	(v_{11}, c_{11})	$\dots$	(v_{1n}, c_{1n})
$\vdots$	$\vdots$	$\ddots$	$\vdots$
v_n	(v_{n1}, c_{n1})	$\dots$	(v_{nn}, c_{nn})

We can record the data of the weights v_{ij} and the network flow capacity c_{ij} simultaneously in the *weight adjacency flow capacity table* (see Table 1.8).

In terms of inventory, let b_i denote the number of units that each particular node has or needs. The convention is that $b_i > 0$ when node v_i has b_i units of product whereas $b_i < 0$ when node b_i requires b_i units of product. The key underlying *network flow assumption* is that

$$\text{The amount shipped out} - \text{The amount shipped in} = \text{The required amount.}$$

The manufacture's aim is to minimize the shipping cost while at the same time meeting the demand for its products.

In order to formulate a suitable linear mathematical programming model, introduce at most $n^2 - n$ variables x_{ij}, for each directed edge $(v_i, v_j) \in E$. Then each decision variable corresponds to a directed edge $(v_i, v_j) \in E$ in the network's *weighted digraph*, with directed edge set E and suitable weights v_{ij}. The idea is that the weights v_{ij} capture particular features of the distribution network.

The manufacture's aim is to minimize the objective function

$$Z = \sum_{(v_i, v_j) \in E} v_{ij} x_{ij}$$

subject to the *network flow constraints*, the *upper bound capacity constraints*, and the *nonnegativity constraints*.

In order to formulate the network flow constraints, denote the existing inventory by positive units and required inventory by negative units. Then, with these conventions

- the network flow constraints are

$$\sum_{(v_i,v_j)\in E} x_{ij} - \sum_{(v_j,v_i)\in E} x_{ji} = b_i \text{ for all } i$$

- the upper bound capacity constraints are

$$x_{ij} \leq c_{ij} \text{ for all } i, j$$

and

- the nonnegativity constraints are

$$x_{ij} \geq 0 \text{ for all } i, j.$$

Summarizing the above discussion, the maximization network flow linear programming problem to be solved may be formulated in the following way.

- Maximize the linear objective function

$$Z = \sum_{(v_i,v_i)\in E} v_{ij}x_{ij}$$

- Subject to the constraints

$$\sum_{(v_i,v_j)\in E} x_{ij} - \sum_{(v_j,v_i)\in E} x_{ji} = b_i \text{ for } i = 1, \ldots, n$$

$$x_{ij} \leq c_{ij} \text{ for all } i, j$$

and

$$x_{ij} \geq 0 \text{ for all } i, j.$$

As explained in Chapter 8, the Simplex Method and its refinements provide a key tool for solving such network flow linear mathematical programming problems.

1.6 Convex Programming Problems

In this section, we describe two representative examples of *nonlinear programming problems*. They fall within the framework of *convex programming problems* which are discussed in Chapter 9.

The first example that we consider, Example 1.22 below, seeks to find optimal production levels so as to maximize a *separable nonlinear profit function* in the presence of resource constraints. As it turns out, when such resource constraints are given by linear inequalities, such nonlinear separable maximization programming problems can be approximated by suitably formulated linear programming problems.

Example 1.22 (Production levels given resource constraints)**.** Here, we consider the case of a manufacturer wanting to determine the optimal production levels for a certain collection of its products given constraints on the resources that it needs to produce these products. The underlying mathematical programming problem is to maximize the manufacturer's total profit.

As a very special case, if each product has a constant unit profit, then the resulting objective function will be linear. However, more generally, consider the case that there are n products and that each product i has a decreasing *price demand curve* $p_i(x_i)$.

Thus, if the unit cost for production and distribution of product i is fixed at a constant cost of c_i, then the manufacturing firm's profit from producing and selling x_i units of product i is given by the nonlinear function

$$P_i(x_i) = x_i p(x_i) - cx_i;$$

thus the total profit function to be maximized takes the form

$$P(x_1, \ldots, x_n) = \sum_{i=1}^{n} (x_i p(x_i) - c_i x_i) .$$

Programming functions of this flavour can be solved, in a systematic way, via techniques for solving *separable programming problems*. Such nonlinear separable programming problems are discussed in Section 9.8. A particularly important case is when the production constraints

$$a_{i1}x_1 + \cdots + a_{in}x_n \leq b_i \text{ for } i = 1, \ldots, m$$

are specified by linear constraints in the decision variables $x_1, \ldots, x_n$.

As explained in Section 9.8, an extremely fruitful technique is to approximate such nonlinear separable programming problems, with linear constraints, by suitably defined linear programming problems. Again, we refer to Section 9.8 for additional details.

The next example that we consider, see Example 1.23 below, is also quite classical. From a historical perspective, it provides strong motivation for the study of quadratic and more general convex programming problems.

Example 1.23 (Optimal Stock Portfolio Selection)**.** Consider the case that an investor is choosing among a collection of n financial products to include in a particular investment portfolio. The investor is concerned with the *expected return* and the expected *risk* that is associated with the potential investments. To this end, a certain quadratic programming model can be used.

For $i = 1, \ldots, n$, let x_i denote the number of units of financial product i that are to be included in the investment portfolio. Let μ_i and σ_{ii}, respectively, be the estimated *mean* and *variance* of the return on each unit of financial product i. Thus, σ_{ii} gives a measure of risk for product i. For $i = 1, \ldots, n, i \neq j$, let σ_{ij} be the *covariance* of the return on one unit of each of product i and product j. In practice, the value for σ_{ij} can be fixed based on historical market prices.

Respectively, the *expected value* $R(x_1, \ldots, x_n)$ and the *variance* $V(x_1, \ldots, x_n)$ of the portfolio's total return are given by

$$R(x_1, \ldots, x_n) = \sum_{j=1}^{n} \mu_j x_j$$

and

$$V(x_1, \ldots, x_n) = \sum_{i=1}^{n} \sum_{j=1}^{n} \sigma_{ij} x_i x_j.$$

Note that $V(x_1, \ldots, x_n)$ gives a measure of the risk that is associated with the entire portfolio. Let L be the minimum expected return. Let P_j be the estimated price for each unit of financial product x_j, for $j = 1, \ldots, n$. Let B be the amount of money that is budgeted for the portfolio.

The programming problem that is of interest is often times referred to as the *Portfolio Selection Problem.* It is formulated in the following way.

- Minimize the objective function

$$V(x_1, \ldots, x_n) = \sum_{i=1}^{n} \sum_{j=1}^{n} \sigma_{ij} x_i x_j$$

- Subject to the constraints

$$\sum_{j=1}^{n} \mu_j x_j \geq L$$

$$\sum_{j=1}^{n} P_j x_j \leq B$$

and

$$x_j \geq 0 \text{ for } j = 1, \ldots, n.$$

Such *quadratic programming problems* fall within the realm of *convex nonlinear programming problems.* (See Chapters 9 and 11.) It turns out that optimal solutions for such programming problems can be found via a variant of the Simplex Method. On the other hand, optimal solutions can also be found using general purpose methods to solve nonlinear convex programming problems. In Chapter 11, we illustrate using Python graphical approaches that employ the matplotlib and numpy packages.

Exercises

1.1 Use the *graphical method* to solve the following maximization linear programming problem.

- Maximize the linear objective function

$$Z = 10x_1 + 20x_2$$

- Subject to the constraints

$$\begin{aligned} -x_1 + 2x_2 &\leq 15 \\ x_1 + x_2 &\leq 12 \\ 5x_1 + 3x_2 &\leq 45 \end{aligned}$$

  and

$$x_1, x_2 \geq 0.$$

1.2 Use the *graphical method* to solve the following minimization linear programming problem.

- Minimize the linear objective function

$$Z = 15x_1 + 20x_2$$

- Subject to the constraints

$$\begin{aligned} x_1 + 2x_2 &\geq 10 \\ 2x_1 - 3x_2 &\leq 6 \\ x_1 + x_2 &\geq 6 \end{aligned}$$

  and

$$x_1, x_2 \geq 0.$$

1.3 An insurance company is introducing new product lines. Respectively, these are special risk insurance and mortgage insurance. The expected profit is \$100/unit on special risk insurance and \$40/unit on mortgages. Management wishes to establish sales quotas for the new product lines to maximize the total expected profit. The work requirements are summarized in Table 1.9.

Table 1.9 Data table for Exercise 1.3

Dept.	Special Risk Work Hours/unit	Mortgage Work Hours/unit	Avail. Work Hours
Underwriting	3	2	2400
Administration	0	1	800
Claims	2	0	1200

(i) Formulate a linear programming model for this problem.
(ii) Use the graphical method to solve this model.

1.4 Consider the following linear programming problem where the value of the constant c_1 is unknown.

- Maximize the linear objective function

$$Z = c_1 x_1 + x_2$$

- Subject to the constraints

$$\begin{aligned} x_1 + x_2 &\le 6 \\ 3x_1 + 6x_2 &\le 30 \end{aligned}$$

and

$$x_1, x_2 \ge 0.$$

Use graphical analysis to determine the optimal solution(s) (x_1, x_2) for the various possible values of c_1, $-\infty < c_1 < \infty$.

1.5 Consider the following linear programming problem where the values of c_1 and c_2 have not yet been determined.

- Maximize the linear objective function

$$Z = c_1 x_1 + c_2 x_2$$

- Subject to the constraints

$$\begin{aligned} 2x_1 + x_2 &\le 11 \\ -2x_1 + 4x_2 &\le 4 \end{aligned}$$

and

$$x_1, x_2 \ge 0.$$

Use graphical analysis to determine the optimal solution(s) $(x_1, x_2) \in \mathbb{R}^2$ for the various possible values of c_1 and c_2.

Table 1.10 Data table for Exercise 1.6

Resource	Q	R	S	Profit per unit
Res. usage per prod. A unit prod.	2	1	3	3
Res. usage per prod. B unit prod.	1	2	3	2
Amount of res. avail.	2	2	4	

1.6 Table 1.10 summarizes the key facts about two products A and B and the resources Q, R and S that are required to produce them.

The aim is to maximize profit.

(i) Formulate a suitable linear programming model.
(ii) Solve this model graphically.

1.7 Consider the case of a communication (or electrical) network that involves a collection of n nodes $v_1, \ldots, v_n$. The communication links could be roads, physical phone or internet lines, electrical power lines or airport terminal hubs. Lines joining the nodes represent direct communication links; the arrows give the directions of those links. Such communication networks can be modelled using *digraphs*.

To get an idea, define, for our purposes here, a *digraph* to consist of a finite collection of vertices $v_1, \ldots, v_n$ together with *directed arcs* that join certain pairs of vertices. A *path* between vertices is then a sequence of arcs that allows one to proceed in a consecutive manner from one vertex to another. The *length* of a path is its number of arcs. A path of length n is called an *n-path*.

A digraph can be completely described by its adjacency matrix. In particular, consider a digraph with vertices $v_1, \ldots, v_n$. The digraph's *adjacency matrix* A is the $n \times n$ matrix with i, j entry given by

$$a_{ij} = \begin{cases} 1 & \text{if there is an arc from } v_i \text{ to } v_j \\ 0 & \text{if there is no arc from } v_i \text{ to } v_j \\ 0 & \text{if } i = j. \end{cases}$$

In a digraph, define the *distance* from vertex v_i to vertex v_j to be the length of the shortest path from v_i to v_j. (If no such path exists then the distance is not defined; we only consider in what follows connected digraphs so in particular there is at least one path from each vertex to another.) The *distance matrix* is the $n \times n$ matrix D which is defined by the condition that its i, j-entry is

$$d_{ij} = \begin{cases} \#\text{ of arcs in a shortest path from vertex } v_i \text{ to vertex } v_j \\ 0 \text{ if } i = j. \end{cases}$$

Consider now the specific case that the matrix

$$A = \begin{bmatrix} 0 & 1 & 1 & 0 & 0 \\ 1 & 0 & 1 & 0 & 0 \\ 0 & 0 & 0 & 1 & 0 \\ 0 & 0 & 0 & 0 & 1 \\ 1 & 0 & 0 & 0 & 0 \end{bmatrix}$$

is the adjacency matrix of a digraph that models a communication network.

(i) Sketch the network.
(ii) Find the digraph's distance matrix.
(iii) Determine the shortest path from node v_2 to node v_5.
(iv) Determine the digraph's *reachability matrix* R which may be defined as the $n \times n$ matrix with i, j entry r_{ij} given by the condition that

$$r_{ij} = \begin{cases} 1 & \text{if there is a path from vertex } v_i \text{ to vertex } v_j \\ 1 & \text{if } i = j \\ 0 & \text{if these is no path from vertex } v_i \text{ to vertex } v_j. \end{cases}$$

Chapter 2
Linear Programming Models—A Collection of Case Study Type Examples

Overview

- A collection of case study type examples
  - resource allocation;
  - nutrition diet problem;
  - mixture blending;
  - agricultural planning; and
  - employee scheduling.

2.1 A Model for Resource Allocation

Here, the setting is of a company that wishes to allocate its resources in such a way that it maximizes its overall measure of performance. A concrete instance is described in Example 2.1.

Example 2.1. A construction company has three work crews and two new long- term contracts. Each contract requires completion of certain tasks at different work sites. Each of the three work crews have specialized areas of expertise and construction efficiencies. Contract 2 requires some work by Crews 1 and 3 but none from Crew 2. Contract 2 requires work by Crew 3 and Crew 2 but none from Crew 1.

Respectively, Crew 1, Crew 2 and Crew 3 have available weekly work capacities of 4, 12 and 18 hours. On the other hand, Contract 1 requires 1 hour of work per task from Crew 1 and 3 hours of work per task from Crew 3. Contract 2 requires 2 hours of work per task from Crew 2 and 2 hours of work per task from Crew 3. Completing a task for Contract 1 yields \$3, 000 profit, whereas completing a task for Contract 2 yields \$5, 000 profit. This data is summarized in Table 2.1.

N. Grieve, *An Introduction to Mathematical Programming and Network Science*,
Springer Undergraduate Texts in Mathematics and Technology,
https://doi.org/10.1007/978-3-032-13330-4_2

Table 2.1 Resource allocation data table for Example 2.1

	Contract 1	Contract 2	Available weekly work time
Crew 1	1	0	4
Crew 2	0	2	12
Crew 3	3	2	18
Profit per completed task	$3, 000	$5, 000	

In order to maximize profits, how many tasks per week from each of the two tasks should the company aim to complete?

In order to solve this optimization problem, via a suitable linear programming problem, let x_1 be the number of tasks completed for Contract 1 per week and let x_2 be the number of tasks completed for Contract 2 per week. Let Z be the total weekly profit from completing each task.

Then the objective function Z is described by

$$Z = 3,000x_1 + 5,000x_2.$$

The constraints that are inherent in the problem are formulated as

$$\begin{aligned} x_1 \qquad\quad &\leq 4 \\ 2x_2 &\leq 12 \\ 3x_1 + 2x_2 &\leq 18 \end{aligned}$$

and

$$x_1, x_2 \geq 0.$$

An optimal solution to the resulting maximization linear programming problem can be found using the graphical methods described in Section 1.4. The objective function Z turns out to be maximized when $x_1 = 2$, $x_2 = 6$ in which case the profit is

$$Z = \$36,000.$$

In other words, in order to maximize profits, the company should aim to complete two tasks towards Contract 1, whereas they should aim to complete six tasks towards Contract 2.

In order to formulate a general model to treat *resource allocation maximization linear programming problems*, such as Example 2.1, the data that is needed to specify such programming problems is described in the following way.

Table 2.2 Resource allocation data table

	Activity			
Resource	1	...	n	Amount of available resource
1	a_{11}	...	a_{1n}	b_1
$\vdots$	$\vdots$	...	$\vdots$	$\vdots$
m	a_{m1}	...	a_{mn}	b_m
Contribution to Z per activity unit	c_1	...	c_n	

First, each such program involves allocation of each of m resources R_i, for $i = 1, \ldots, m$, to each of n activities A_j, for $j = 1, \ldots, n$. There are given amounts of available resources b_i for $i = 1, \ldots, m$.

The objective function

$$Z = c_1x_1 + \cdots + c_nx_n$$

gives an overall measure of performance. Here, the decision variables x_j, for $j = 1, \ldots, n$, represent activity levels and the coefficients c_j, for $j = 1, \ldots, n$, represent performance increase for each unit increase in level of activity and amount of available resource b_i, for $i = 1, \ldots, m$.

Letting a_{ij}, for $i = 1, \ldots, m$ and $j = 1, \ldots, n$, denote the amount of resource i that is consumed by each unit of activity j, the data needed to specify a resource allocation maximization linear programming problem is summarized in Table 2.2.

The *input constants* c_j, b_j and a_{ij}, for $i = 1, \ldots, m$ and $j = 1, \ldots, n$, are the *parameters* of the model.

Given such a resource allocation table, see Table 2.2, the corresponding *resource allocation maximization* linear programming problem is formulated as follows.

- Maximize the linear objective function

$$Z = c_1x_1 + \cdots + c_nx_n$$

- Subject to the constraints

$$\begin{array}{ccccc} a_{11}x_1 & + \ldots + & a_{1n}x_n & \leq & b_1 \\ \vdots & \vdots & \vdots & & \vdots \\ a_{m1}x_1 & + \ldots + & a_{mn}x_n & \leq & b_m \end{array}$$

and

$$x_1, \ldots, x_n \geq 0.$$

2.2 A Model for Nutrition Diet Problems

Consider the problem of calculating the minimum cost required to provide an individual with the recommended amount of daily energy E (with units kcal), protein P (with units g) and calcium C (with units mg). A concrete case is described in Example 2.2.

Example 2.2. One guide for healthy breakfasts is that they should include a mixture of protein, fruits and complex carbohydrates. For instance, such a breakfast might contain a collection of eggs, oatmeal, yogurt, bread, nuts and fruits. It might also have at least 400 calories, at least 30 grams of protein and at least 450 milligrams of calcium.

The aim is to minimize the cost required in order to meet the given dietary constraints and subject to a given collection of available food sources, prices and nutrition values. For instance, suppose that such data is summarized in Table 2.3.

Respectively, let $x_1, \ldots, x_6$ denote the number of daily servings of eggs, oatmeal, yogurt, bread, nuts and fruits. Then the total daily breakfast cost is described by the objective function

$$Z = .50x_1 + .25x_2 + .90x_3 + .25x_4 + 3.25x_5 + 1.25x_6.$$

Assuming that there exist no restrictions on the number of servings of any particular food source, the aim is to minimize Z subject to the constraints

$$\begin{aligned}
.16x_1 + .18x_2 + .15x_3 + .16x_4 + .125x_5 + .170x_6 &\geq .4 \\
13x_1 + 2.4x_2 + 10x_3 + 9x_4 + 20x_5 + 3x_6 &\geq 30 \\
50x_1 + 150x_2 + 400x_3 + 26x_4 + 24x_5 + 27x_6 &\geq 450
\end{aligned}$$

and

$$x_1, \ldots, x_6 \geq 0.$$

Table 2.3 Nutrition diet data table for Example 2.2

Food Source	Energy per serving (kcal)	Protein per serving (g)	Calcium per serving (mg)	Price per serving (dollars)
two eggs	.16	13	50	.50
oatmeal	.18	2.4	150	.25
yogurt	.15	10	400	.90
bread	.16	9	26	.25
nuts	.125	20	24	3.25
fruits	.170	3	27	1.25

Table 2.4 Nutrition diet data table

Food Source	Energy per serving (kcal)	Protein per serving (g)	Calcium per serving (mg)	Price per serving (dollars)
F_1	e_1	p_1	c_1	d_1
$\vdots$	$\vdots$	$\vdots$	$\vdots$	$\vdots$
F_n	e_n	p_n	c_n	d_n

As follows from the Simplex Method, see for instance Example 6.8, an optimal solution for this programming problem turns out to be

$$Z = \$1.30$$

when $x_2 = 2.54$, $x_4 = 2.66$ and $x_1 = x_3 = x_5 = x_6 = 0$.

Especially, these calculations indicate that a breakfast which consists entirely of oatmeal and bread will be of minimal cost while at the same time will meet all of the nutritional requirements.

In order to formulate a general model for nutrition diet problems, such as Example 2.2, consider the data of n food sources with nutrient and cost data summarized in Table 2.4.

Let x_i denote the number of daily servings of food source i, for $i = 1, \ldots, n$. Then the total daily food cost is described by the objective function

$$Z = d_1 x_1 + \cdots + d_n x_n.$$

On the other hand, under the assumption that there exist no additional restrictions on the number of servings of any particular food source, and assuming that partial servings of each particular food source are possible, then a solution to the diet problem can be expressed as a solution to the following minimization linear programming problem.

- Minimize the objective function

$$Z = x_1 d_1 + \cdots + x_n d_n$$

- Subject to the constraints

$$\begin{aligned} e_1 x_1 + \ldots + e_n x_n &\geq E \\ p_1 x_1 + \ldots + p_n x_n &\geq P \\ c_1 x_1 + \ldots + c_n x_n &\geq C \end{aligned}$$

  and

$$x_1, \ldots, x_n \geq 0.$$

Here, E is the recommended amount of daily energy (with units kcal), P is the recommended amount of daily protein (with units g) and C is the recommended amount of daily calcium (with units mg).

As a slight variant to this problem, if we wish to place constraints on the number of servings s_i of food source F_i that may be used in any given particular day, then the solution to the diet problem with serving restrictions can be expressed as a solution to the following minimization linear programming problem.

- Minimize the objective function

$$Z = x_1 d_1 + \cdots + x_n d_n$$

- Subject to the constraints

$$\begin{aligned} e_1 x_1 + \ldots + e_n x_n &\geq E \\ p_1 x_1 + \ldots + p_n x_n &\geq P \\ c_1 x_1 + \ldots + c_n x_n &\geq C \end{aligned}$$

and

$$0 \leq x_1 \leq s_1, \ldots, 0 \leq x_n \leq s_n.$$

2.3 A Model for Mixture Blending with Production Levels and Selling Strategies

Oil refineries produce various types of raw gasoline and monitors the performance number and vapour pressure. The aim is to find an optimal mixture blend given various production levels and selling strategies. A sample situation is described in Example 2.3.

Example 2.3. Suppose that an oil refinery produces four types of raw gasoline. Two important features of each type are its antiknock properties, which is given by its *performance number* and its volatility, which is given by its *vapour pressure number*. Suppose that these two characteristics together with the production levels in barrels per day are described in Table 2.5.

Table 2.5 Performance and vapour pressure data table for Example 2.3

Type of Gasoline	Performance Number	Vapour Pressure	Barrels produced per day
Type 1	115	6	3, 980
Type 2	105	5	6, 728
Type 3	119	6	4, 101
Type 4	101	5	2, 369

Table 2.6 Production data table for Example 2.3

	Performance Number	Vapour Pressure	Price per Barrel
Variant 1	≥ 110	≤ 6	\$64.50
Variant 2	≥ 95	≤ 6	\$59.91

The gasoline will be sold in two types of blended form. Quality standards impose restrictions on the respective performance number and vapour pressure. These restrictions and the respective selling prices are described in Table 2.6.

The performance number and volatility of each variant is given by the weighted averages of the performance number and volatility of its constituents. The aim is to devise a production plan that will yield the largest revenue when 5537 barrels of Variant 1 and 6572 barrels of Variant 2 are produced.

To model this situation via a suitable linear programming problem let x_{ij}, for $i = 1, \ldots, 4$ and $j = 1, 2$, denote the number of barrels of Type i gasoline that is used to make Variant j. Respectively, let PN_1 and PN_2 and VP_1 and VP_2 be the performance number and vapour pressure for Variant 1 and Variant 2. Then

$$\text{PN}_1 = \frac{115x_{11} + 105x_{21} + 119x_{31} + 101x_{41}}{5537}$$

$$\text{PN}_2 = \frac{115x_{12} + 105x_{22} + 119x_{32} + 101x_{42}}{6572}$$

$$\text{VP}_1 = \frac{6x_{11} + 5x_{21} + 6x_{31} + 5x_{41}}{5537}$$

and

$$\text{VP}_2 = \frac{6x_{12} + 5x_{22} + 6x_{32} + 5x_{42}}{6572}.$$

The resulting maximization linear programming problem is expressed as follows.

- Maximize the revenue function

$$R = (64.50)(x_{11} + x_{21} + x_{31} + x_{41}) + (59.91)(x_{12} + x_{22} + x_{32} + x_{42})$$

- Subject to the constraints

$$\begin{array}{rcrcrcrcl}
115x_{11} & + & 105x_{21} & + & 119x_{31} & + & 101x_{41} & \geq & 609070 \\
115x_{12} & + & 105x_{22} & + & 119x_{32} & + & 101x_{42} & \geq & 624340 \\
6x_{11} & + & 5x_{21} & + & 6x_{31} & + & 5x_{41} & \leq & 33222 \\
6x_{12} & + & 5x_{22} & + & 6x_{32} & + & 5x_{42} & \leq & 39432 \\
x_{11} & + & x_{21} & + & x_{31} & + & x_{41} & = & 5537 \\
x_{12} & + & x_{22} & + & x_{32} & + & x_{42} & = & 6572
\end{array}$$

$$x_{11} + x_{12} \leq 3980$$
$$x_{21} + x_{22} \leq 6728$$
$$x_{31} + x_{32} \leq 4101$$
$$x_{41} + x_{42} \leq 2369$$

and

$$x_{11}, \ldots, x_{42} \geq 0.$$

It turns out, see, for example, Example 6.9, as an application of the Simplex Method, that an optimal solution to this linear programming problem is achieved when $x_{11} = 0$, $x_{21} = 1436$, $x_{31} = 4101$, $x_{41} = 0$, $x_{12} = 0$, $x_{22} = 4203$, $x_{23} = 0$, $x_{42} = 2369$. The optimal value of the objective function is

$$R = \$750865.02.$$

In particular, the conclusion is that maximum revenue is achieved when Type 2 and Type 3 gasoline are used to make Variant 1, whereas Type 2 and Type 3 gasoline are used to make Variant 2.

In order to formulate a general model, for *mixture blending*, suppose that the oil refinery produces m types of raw gasoline, Type 1, ..., Type m, and monitors the *performance number* PN and *vapour pressure* VP of each particular type. Recall that, respectively, the PN and VP of each particular type of gasoline give a measure of its antiknocking properties and its volatility. Suppose that the PN and VP together with the production levels, in barrels per day, for each particular type of raw gasoline is given in Table 2.7.

Suppose that the company sells the gasoline in a collection of n blended variants, denoted as Variant 1, ..., Variant n, and that the intended price per barrel for selling each particular variant i is given by d_i dollars and that the company wishes to make B_j barrels of Variant j, for $j = 1, \ldots, n$. Further, suppose that the quality control standards impose certain requirements on the performance number and vapour pressure of each particular variant. The relevant data is summarized in Table 2.8.

Suppose that the PN and VP of each variant is given by the weighted average of the PN and VP of each of the types that it contains. The refinery aims to determine the mixing strategy that will yield the largest revenue.

Table 2.7 Performance and vapour pressure data table

Type of Gasoline	PN	VP	Barrels produced per day
Type 1	q_1	v_1	b_1
$\vdots$	$\vdots$	$\vdots$	$\vdots$
Type m	q_m	v_m	b_m

Table 2.8 Production data table

Variant	PN	VP	Revenue per barrel	Number of barrels to produce
Variant 1	$\geq p_1$	$\leq w_1$	d_1	B_1
$\vdots$	$\vdots$	$\vdots$	$\vdots$	$\vdots$
Variant n	$\geq p_n$	$\leq w_n$	d_n	B_n

To model this situation via a linear programming problem, let x_{ij}, for $i = 1, \dots, m$ and $j = 1, \dots, n$, denote the number of barrels of Type i gasoline that is used to make Variant j. Respectively, let PN_j be the performance number and VP_j be the vapour pressure of variant j, for $j = 1, \dots, n$. Then

$$\text{PN}_j = \frac{q_1 x_{1j} + \cdots + q_m x_{mj}}{x_{1j} + \cdots + x_{mj}} = \frac{q_1 x_{1j} + \cdots + q_m x_{mj}}{B_j}$$

and

$$\text{VP}_j = \frac{v_1 x_{1j} + \cdots + v_m x_{mj}}{x_{1j} + \cdots + x_{mj}} = \frac{v_1 x_{1j} + \cdots + v_m x_{mj}}{B_j}.$$

Finally, the total revenue is given by

$$R = (x_{11} + \cdots + x_{m1})d_1 + \cdots + (x_{1n} + \cdots + x_{mn})d_n.$$

A solution to the revenue maximization problem is then given by the following maximization linear programming problem.

- Maximize the objective function

$$R = (x_{11} + \cdots + x_{m1})d_1 + \cdots + (x_{1n} + \cdots + x_{mn})d_n$$

- Subject to the constraints

$$q_1 x_{1j} + \cdots + q_m x_{mj} - (x_{1j} + \cdots + x_{mj})p_j \geq 0 \text{ for } j = 1, \dots, n$$

$$v_1 x_{1j} + \cdots + v_m x_{mj} - (x_{1j} + \cdots + x_{mj})w_j \leq 0 \text{ for } j = 1, \dots, n$$

$$x_{i1} + \cdots + x_{in} \leq b_i \text{ for } i = 1, \dots, m$$

$$x_{1j} + \cdots + x_{mj} = B_j \text{ for } j = 1, \dots, n$$

and

$$x_{ij} \geq 0 \text{ for } i = 1, \dots, m \text{ and } j = 1, \dots, n.$$

2.4 A Model for Agricultural Planning

Consider the case of a collection of agricultural fields with usable land and available water allocation. The aim is to maximize the net return given certain logistical and production quota constraints. Example 2.4 gives a concrete example.

Example 2.4. Three community gardens with usable land and water allocation are summarized in Table 2.9.

Suppose that, based on historical data, production quota limits and net return, for selling at the local Farmer's market, the relevant data for 4 vegetables is summarized in Table 2.10.

Let x_{ij} for $i = 1, 2, 3, 4$ and $j = 1, 2, 3$ represent the amount, in m^2, of vegetable i planted in garden j. The objective function that is to be maximized is described as

$$Z = 14(x_{11} + x_{12} + x_{13}) + 3(x_{21} + x_{22} + x_{23}) \\ + 20(x_{31} + x_{32} + x_{33}) + 16(x_{41} + x_{42} + x_{43}).$$

There are several constraints to consider. Especially

- usable land constraints for each of the three gardens

$$\begin{aligned} x_{11} + x_{21} + x_{31} + x_{41} &\leq 37 \\ x_{12} + x_{22} + x_{32} + x_{42} &\leq 45 \\ x_{13} + x_{23} + x_{33} + x_{43} &\leq 65 \end{aligned}$$

Table 2.9 Land allocation table for Example 2.4

Community Garden	Usable Land (m^2)	Water Allocation ($L//m^2$)
Garden 1	37	300
Garden 2	45	350
Garden 3	65	700

Table 2.10 Agricultural planning table for Example 2.4

Vegetable	Max Quota (m^2)	Water Use (L / /m^2)	Net Return ($ / /m^2)
Carrots	20	25	$14
Potatoes	30	35	$3
Lettuce	15	28	$20
Beets	12	26	$16

- water allocation constraints for each of the three gardens

$$
\begin{aligned}
25x_{11} + 35x_{21} + 28x_{31} + 26x_{41} &\le 300 \\
25x_{12} + 35x_{22} + 28x_{32} + 26x_{42} &\le 350 \\
25x_{13} + 35x_{23} + 28x_{33} + 26x_{43} &\le 700
\end{aligned}
$$

- total area constraints for each vegetable

$$
\begin{aligned}
x_{11} + x_{12} + x_{13} &\le 20 \\
x_{21} + x_{22} + x_{23} &\le 30 \\
x_{31} + x_{32} + x_{33} &\le 15 \\
x_{41} + x_{42} + x_{43} &\le 12
\end{aligned}
$$

and

- equal proportion of land planted constraints

$$
\frac{x_{11} + x_{21} + x_{31} + x_{41}}{37} = \frac{x_{12} + x_{22} + x_{32} + x_{42}}{45} = \frac{x_{13} + x_{23} + x_{33} + x_{43}}{65}.
$$

- Finally, there are the nonnegativity constraints

$$
x_{ij} \ge 0 \text{ for } i = 1, 2, 3, 4 \text{ and } j = 1, 2, 3.
$$

The equal proportion of land planted constraints serve to ensure an equitable out put of each of the three gardens.

It turns out, see Example 6.10, that an optimal solution is

$$
Z = 754.27.
$$

This occurs when $x_{11} = 0, x_{12} = 14, x_{13} = 4.73, x_{21} = 0, x_{22} = 0, x_{23} = 0, x_{31} = 0, x_{32} = 0, x_{33} = 15, x_{41} = 11.51, x_{42} = 0, x_{43} = .49$.

The conclusion is thus that the net return is maximized by planting carrots in Gardens 2 and 3, lettuce in Garden 3 and beets in Gardens 1 and 3. Potatoes should not be planted at all.

In order to formulate a more general model, for *agriculture planning*, consider a collection of n agricultural fields with usable land and available water allocation summarized as in Table 2.11.

Suppose that based on past historical data and on production quota limits, the relevant crop data for m suitable crops to be planted in these fields is summarized in Table 2.12.

Now, to ensure a good crop yield in each field, suppose that the same proportion of available irrigable land will be planted in each field. On the other hand, suppose that there is no particular restriction on the combination of the crops to be grown at each particular field.

Table 2.11 Land allocation table

Agricultural Field	Usable Land (m^2)	Water Allocation ($L//m^2$)
Field 1	ℓ_1	a_1
$\vdots$	$\vdots$	$\vdots$
Field n	ℓ_n	a_n

Table 2.12 Agricultural planning table

Crop	Max Quota (m^2)	Water Use ($L\ /\ /m^2$)	Net Return $ / $/m^2$
Crop 1	q_1	u_1	d_1
$\vdots$	$\vdots$	$\vdots$	$\vdots$
Crop m	q_m	u_m	d_m

The goal is then to decide how many acres to devote to each particular crop at each respective field while at the same time satisfying each of the given restrictions. Especially, the aim is to maximize the total net return when considering all n fields together.

We need to decide on the number of acres to devote to each of the m crops at each of the n fields. There are mn decision variables x_{ij}, for $i = 1, \ldots, m$ and $j = 1, \ldots, n$. Here, x_{ij} represents the amount of crop i that is allocated to field j. The objective function is

$$Z = \sum_{j=1}^{n} \sum_{i=1}^{m} d_i x_{ij}.$$

Thus, Z represents the total net return and the aim is to maximize the objective function Z subject to the given *agricultural planning constraints*. These take the following form.

- Usable land for each field constraints

$$x_{1j} + \cdots + x_{mj} \leq \ell_j \text{ for } j = 1, \ldots, n$$

- water allocation for each field constraints

$$u_1 x_{1j} + \cdots + u_m x_{mj} \leq a_j \text{ for } j = 1, \ldots, n$$

- total acreage for each crop constraints

$$x_{i1} + \cdots + x_{in} \leq q_i \text{ for } i = 1, \ldots, m$$

- equal proportion of land planted constraints

$$\frac{x_{11} + \cdots + x_{m1}}{\ell_1} = \cdots = \frac{x_{1n} + \cdots + x_{mn}}{\ell_n}$$

and
- nonnegativity constraints

$$x_{ij} \geq 0 \text{ for } i = 1, \ldots, m \text{ and } j = 1, \ldots, n.$$

As highlighted in Example 2.4, the equal proportion of land constraints serve to ensure equality of output by each agricultural field.

A solution to the question of maximizing the total net return subject to the given constraints can thus be found upon solving the following maximization linear program.

- Maximize the objective function

$$Z = \sum_{j=1}^{n} \sum_{i=1}^{m} d_i x_{ij}$$

- Subject to the constraints

$$x_{1j} + \cdots + x_{mj} \leq \ell_j \text{ for } j = 1, \ldots, n$$

$$u_1 x_{1j} + \cdots + u_m x_{mj} \leq a_j \text{ for } j = 1, \ldots, n$$

$$x_{i1} + \cdots + x_{in} \leq q_i \text{ for } i = 1, \ldots, m$$

$$\frac{x_{11} + \cdots + x_{m1}}{\ell_1} = \cdots = \frac{x_{1n} + \cdots + x_{mn}}{\ell_n}$$

and

$$x_{ij} \geq 0 \text{ for } i = 1, \ldots, m \text{ and } j = 1, \ldots, n.$$

2.5 A Model for Employee Scheduling Subject to Minimum Staffing Requirements

Consider the problem of scheduling employees subject to minimum staffing requirements. The idea is to ensure that a satisfactory level of service is achieved. A concrete instance of this problem is described in Example 2.5 below.

Example 2.5. Suppose that a hospital needs to schedule its support staff employees subject to minimum staffing requirements in order to ensure that a satisfactory level of service is achieved. The provisions of the collective agreement are such that each employee work an 8-hour shift 5 days a week.

The authorized shifts together with the relevant staffing requirement data are respectively summarized in Tables 2.13 and 2.14.

The x's in the middle portion of the table show the hours covered by the respective shifts. Note that the wages differ by shift and further that the bottom row of the table gives the daily compensation rate for each employee for each given shift.

The aim of the hospital administrator is to determine the optimal number of employees that should be assigned to the respective daily shifts in order to minimize the total employee compensation costs. In order to formulate a suitable linear programming problem, there are five decision variables. Indeed, let x_j denote the number of employees assigned to shift j for $j = 1, \ldots, 5$.

Table 2.13 Table of authorized shifts for Example 2.5

Shift 1:	$06:00 - 14:00$
Shift 2:	$08:00 - 16:00$
Shift 3:	$12:00 - 20:00$
Shift 4:	$16:00 - 00:00$
Shift 5:	$22:00 - 06:00$

Table 2.14 Table of staffing requirements for Example 2.5

Shift	1	2	3	4	5	Minimum staffing levels
Time period						
06:00–08:00	x					48
08:00–10:00	x	x				79
10:00–12:00	x	x				65
12:00–14:00	x	x	x			87
14:00–16:00		x	x			64
16:00–18:00			x	x		53
18:00–20:00			x	x		62
20:00–22:00				x		72
22:00–00:00				x	x	25
00:00–06:00					x	15
Average daily cost per employee ($)	190	205	215	225	235	

The objective is to minimize the total cost of the employees assigned to each of the five shifts. This is achieved via the following minimization linear program.

- Minimize the linear objective function

$$Z = 190x_1 + 205x_2 + 215x_3 + 225x_4 + 235x_5$$

- Subject to the constraints

$$\begin{aligned}
x_1 & \geq 48 \\
x_1 + x_2 & \geq 79 \\
x_1 + x_2 & \geq 65 \\
x_1 + x_2 + x_3 & \geq 87 \\
x_2 + x_3 & \geq 64 \\
x_3 + x_4 & \geq 53 \\
x_3 + x_4 & \geq 62 \\
x_4 & \geq 72 \\
x_4 + x_5 & \geq 25 \\
x_5 & \geq 15
\end{aligned}$$

and

$$x_1, \ldots, x_5 \geq 0.$$

(Note that some of the above constraints are redundant.)

It turns out, see Example 6.11, that the optimal value for the objective function is

$$Z = 41965.$$

This is achieved when $x_1 = 48, x_2 = 64, x_3 = 0, x_4 = 72$ and $x_5 = 15$. Significantly, the daily employee cost is minimized when no employee is assigned to shift three.

In order to formulate a general model, suppose that the provisions of the collective agreement are such that each employee work an 8-hour shift 5 days a week and that the authorized shifts are described in Table 2.15.

The relevant shift scheduling data is then summarized in Table 2.16.

Table 2.15 Table of authorized shifts

Shift 1:	06 : 00 – 14 : 00
Shift 2:	08 : 00 – 16 : 00
Shift 3:	12 : 00 – 20 : 00
Shift 4:	16 : 00 – 00 : 00
Shift 5:	22 : 00 – 06 : 00

Table 2.16 Table of staffing requirements

Shift	1	2	3	4	5	Minimum staffing levels
Time period						
06:00–08:00	x					s_1
08:00–10:00	x	x				s_2
10:00–12:00	x	x				s_3
12:00–14:00	x	x	x			s_4
14:00–16:00		x	x			s_5
16:00–18:00			x	x		s_6
18:00–20:00			x	x		s_7
20:00–22:00				x		s_8
22:00–00:00				x	x	s_9
00:00–06:00					x	s_{10}
Average daily cost per employee ($)	d_1	d_2	d_3	d_4	d_5	

In that table, as before, the x's in the middle portion of the table show the hours covered by the respective shifts. The wages differ by shift. The bottom row of the table gives the daily compensation rate d_i dollars for each employee for each given shift $j = 1, \ldots, 5$. The s_i, for $i = 1, \ldots, 10$, give the minimum required staffing numbers for each of the 10 given time periods.

The *employee scheduling optimization problem*, seeks to determine the optimal number of employees that should be assigned to the respective daily shifts in order to minimize the total employee compensation costs. The constraints are the given minimum staffing requirements that are specified by the rightmost column.

To formulate a suitable linear programming model, let x_j denote, for $j = 1, \ldots, 5$, the number of employees assigned to shift j. The objective is to minimize the total cost of the employees assigned to each of the five shifts. A solution to the employee scheduling optimization problem, can be found by solving the following linear minimization programming problem.

- Minimize the linear objective function

$$Z = d_1x_1 + d_2x_2 + d_3x_3 + d_4x_4 + d_5x_5$$

• Subject to the constraints

$$
\begin{array}{lr}
x_1 & \geq s_1 \\
x_1 + x_2 & \geq s_2 \\
x_1 + x_2 & \geq s_3 \\
x_1 + x_2 + x_3 & \geq s_4 \\
\quad x_2 + x_3 & \geq s_5 \\
\quad\quad x_3 + x_4 & \geq s_6 \\
\quad\quad x_3 + x_4 & \geq s_7 \\
\quad\quad\quad x_4 & \geq s_8 \\
\quad\quad\quad x_4 + x_5 & \geq s_9 \\
\quad\quad\quad\quad x_5 & \geq s_{10}
\end{array}
$$

and

$$x_1, \ldots, x_5 \geq 0.$$

Exercises

2.1 A butcher shop makes its meat loaf from two types of meat. The Type 1 meat contains 80% meat and 20% fat and costs \$3.20/lb. The Type 2 meat contains 68% meat and 32% fat and costs \$2.80/lb. How much of each kind of meat should the shop use in each pound of meat loaf in order to minimize its cost and keep the fat content of the meat loaf to no more than 25%?

2.2 An airline passenger wants to pack 5 items, but together they exceed the airline's 50lb weight limit. To assist with the decision, the passenger assigns a preference value to each item. This is summarized in Table 2.17.

Formulate a suitable linear programming problem whose solution will provide insight as to which items should be taken in order to maximize the total preference value without exceeding the airline's checked baggage weight restrictions.

2.3 A 24-hour supermarket has minimal requirements for cashiers. The relevant data is given in Table 2.18.

In Table 2.18, the time period 1 follows immediately after the time period 6. Also, a cashier works eight consecutive hours, starting at the beginning of one of the six

Table 2.17 Data table for Exercise 2.2

Item	1	2	3	4	5
Weight (lb)	28	13	5	15	7
Value	100	60	70	15	15

Table 2.18 Data table for Exercise 2.3

Period	1	2	3	4	5	6
Time of day (24 hr clock)	3–7	7–11	11–15	15–19	19–23	23–3
Minimum no.	7	20	14	20	10	5

periods. Determine a suitable linear programming problem whose solution will give a daily employee worksheet which satisfies the requirement with the least number of personnel.

2.4 A manufacturer has 1200 boxes of product in stock at one factory and another 1000 boxes at a second factory. The manufacturer has orders for this product from three different retailers in quantities of 1000, 700 and 500 boxes, respectively.

The unit shipping costs (in cents per box) from the factories to the retailers are described in Table 2.19.

Determine a suitable linear programming problem whose solution will give a minimum cost shipping schedule for satisfying all demands from current inventory.

2.5 A 400m medley swim relay involves four different swimmers who each successively swim 100m of the backstroke, breast stroke, butterfly and freestyle. A coach has six swimmers whose expected times (in seconds) in the individual events are given in Table 2.20.

Determine a suitable linear programming problem whose solution will give insight as to how the swim coach should assign swimmers to the relay so as to minimize the sum of their times.

Table 2.19 Data table for Exercise 2.4

	Retailer 1	Retailer 2	Retailer 3
Factory 1	14	13	11
Factory 2	13	13	12

Table 2.20 Data table for Exercise 2.5

	Event 1 (Backstroke)	Event 2 (Breaststroke)	Event 3 (Butterfly)	Event 4 (Freestyle)
Swimmer 1	53.78	59.55	51.69	48.71
Swimmer 2	53.90	60.01	51.75	48.34
Swimmer 3	53.85	59.58	51.68	48.43
Swimmer 4	54.01	59.51	51.85	48.35
Swimmer 5	53.95	60.10	52.05	48.37
Swimmer 6	53.81	59.99	51.59	48.42

Table 2.21 Data table for Exercise 2.6

	Orange juice %	Grapefruit juice %	Cranberry juice %	Supply litres	Cost $ / /litre
Drink 1	40	40	0	200	$1.50
Drink 2	5	10	20	400	$0.75
Drink 3	100	0	0	100	$2.00
Drink 4	0	100	0	50	$1.75
Drink 5	0	0	0	800	$0.25

2.6 Suppose that a wedding caterer must prepare from five in stock fruit drinks, 500 litres of a punch that contains at least 20% orange juice, 10% grapefruit juice and 5% cranberry juice. The in stock inventory data is summarized in Table 2.21.

Formulate a minimization linear programming problem which is such that an optimal solution will determine the minimum cost composition of the fruit punch.

Chapter 3
Towards a Theory for Mathematical Programming Problems

Overview

- First examples of the main geometric method.
- Introduction to canonical form linear programming problems.
- Selected convex geometric considerations.
- Theoretical justification of the main geometric method for solving linear programming problems.

3.1 First Examples of the Main Geometric Method

In this section, we revisit again the graphical method, which is described in Section 1.4. Its theoretical justification follows from Theorem 3.24. In order to formulate Theorem 3.24 precisely, we need to introduce some first concepts from convex geometry. This is the aim of Section 3.6. Revisiting the graphical method here, see Examples 3.1 and 3.2, prior to formulating these more abstract convex geometric notions is useful. It illustrates the essentially exhaustive search method that the Simplex Method, see Chapter 5, ultimately refines.

Example 3.1. Consider the following maximization linear program.

- Maximize the linear objective function

$$P(x, y) = 600x + 400y$$

- Subject to the constraints

$$\begin{aligned} 2x + \ y &\leq 8 \\ x + 2y &\leq 10 \end{aligned}$$

N. Grieve, *An Introduction to Mathematical Programming and Network Science*,
Springer Undergraduate Texts in Mathematics and Technology,
https://doi.org/10.1007/978-3-032-13330-4_3

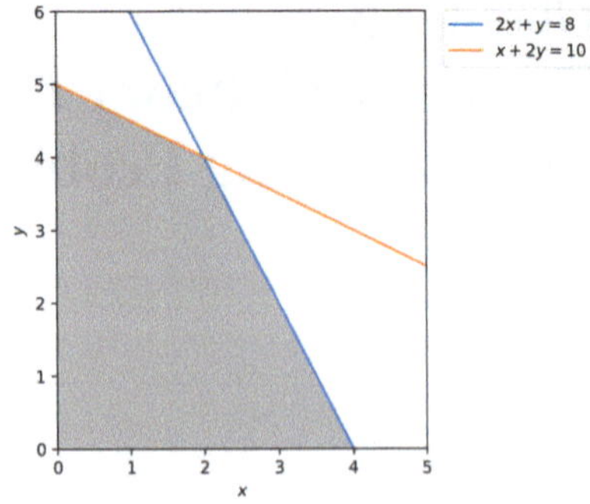

Fig. 3.1 The feasible region for Example 3.1

and

$$x, y \geq 0.$$

The feasible region is described as the shaded region that is given in Figure 3.1.

The extreme points of the constraint set are

$$(0, 0), (4, 0), (0, 5) \text{ and } (2, 4) \in \mathbb{R}^2.$$

The given linear programming problem can be solved by evaluating the objective function $P(x, y)$ at each of these extreme points and then by noting which evaluations yield maximums. (See Table 3.1.)

Thus, $P(x, y)$ is maximized at the extreme point $(2, 4) \in \mathbb{R}^2$.

Example 3.2. Consider the following minimization linear program.

- Minimize the linear objective function

$$C(x, y) = 3500x + 4500y$$

- Subject to the constraints

$$\begin{aligned} 20x + 40y &\geq 1000 \\ 25x + 20y &\geq \ 800 \end{aligned}$$

and

$$x, y \geq 0.$$

Table 3.1 The extreme points and objective function values for Example 3.1

Extreme point $(x, y) \in \mathbb{R}^2$	Value of $P(x, y)$ at extreme point
(0, 0)	0
(4, 0)	2400
(0, 5)	2000
(2, 4)	2800

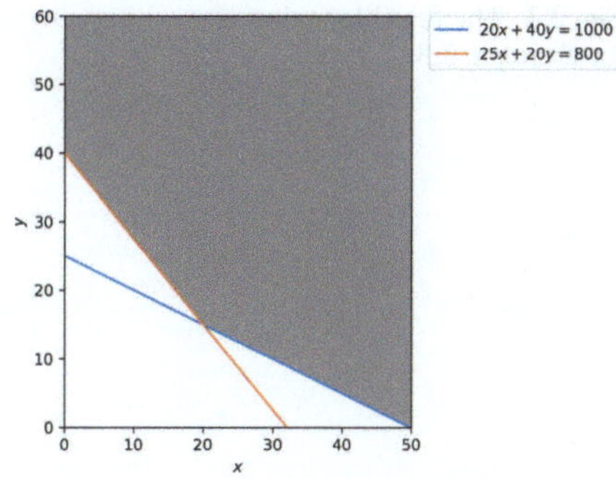

Fig. 3.2 The feasible region for Example 3.2

The feasible region is described as the shaded region depicted in Figure 3.2.

It is unbounded. However, the objective function $C(x, y)$ is bounded from below. Indeed, increasing the x and/or y values over the feasible region evidently does not decrease the value of $C(x, y)$. The extreme points of the feasible region are

$$(50, 0), (0, 40) \text{ and } (20, 15) \in \mathbb{R}^2.$$

Hence a solution to the given minimization linear programming problem can be found by evaluating $C(x, y)$ at each extreme point and the observing which evaluation(s) yield minimum value(s). This is described in Table 3.2.

Therefore, $C(x, y)$ is minimized at the extreme point $(20, 15) \in \mathbb{R}^2$.

As mentioned, the graphical method employed in Examples 3.1 and 3.2 should be seen as special instances of the conclusion provided by Theorem 3.24.

3.2 Introduction to Sensitivity Analysis

The concept of *sensitivity analysis* is an important aspect of mathematical programming. Briefly, the idea is to investigate the effect on the optimal solution of a given mathematical program upon making changes to the input parameters. For instance, certain parameters of a given mathematical programming problem, such as resource amounts, may represent managerial decisions. Others, such as commodity prices, may depend on market conditions.

Table 3.2 The extreme points and objective function values for Example 3.2

Extreme point $(x, y) \in \mathbb{R}^2$	Value of $C(x, y)$ at extreme point
(50, 0)	175000
(0, 40)	180000
(20, 15)	137500

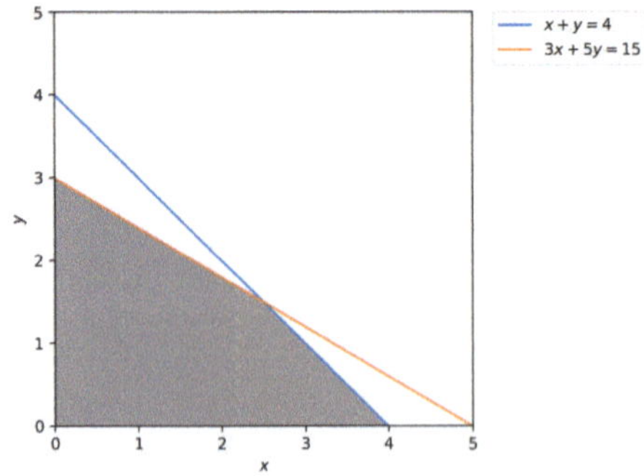

Fig. 3.3 The feasible region for Example 3.3

Sensitivity analysis applied to a given programming problem involves studying the extent to which an optimal solution relies on the given input parameters. In Example 5.9, we revisit the concept of sensitivity analysis from the view point of the Simplex Method. Here, see Example 3.3 below, we make qualitative observations. They illustrate the manner in which sensitivity analysis interacts with a given mathematical program's feasible region.

Example 3.3. Suppose that a given linear programming problem has feasible region described by the inequalities

$$\begin{aligned} x + y &\leq 4 \\ 3x + 5y &\leq 15 \end{aligned}$$

and

$$x, y \geq 0$$

and objective function

$$f(x, y) = 3x + 4y.$$

The feasible region's corner points are $(0, 0)$, $(0, 3)$, $\left(\frac{5}{2}, \frac{3}{2}\right)$ and $(4, 0)$.

The objective function $f(x, y)$ is maximized over the feasible region when $x = \frac{5}{2}$ and $y = \frac{3}{2}$; its maximal value is 13.5. The feasible region is pictured in Figure 3.3.

Consider a small perturbation of the feasible region. For example, consider replacing the feasible region's constraints by

$$\begin{aligned} 5x + 3y &\leq 15 \\ 2x + 6y &\leq 12 \end{aligned}$$

and

$$x, y \geq 0.$$

This new feasible region is pictured in Figure 3.4

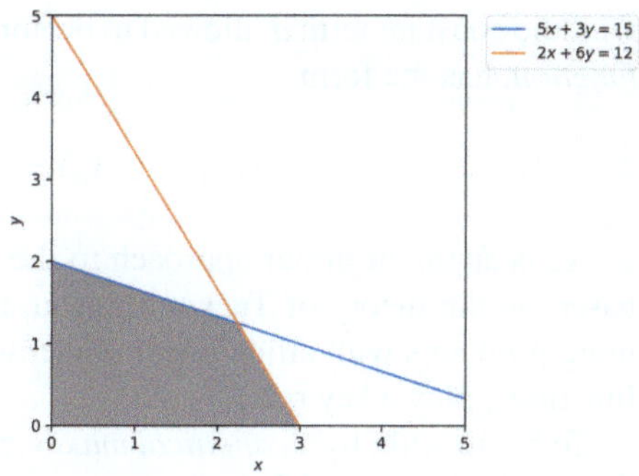

Fig. 3.4 The perturbed feasible region for Example 3.3

The corner points are $(0, 0)$, $(0, 2)$, $\left(\frac{9}{4}, \frac{5}{4}\right)$, $(3, 0)$ and the objective function

$$f(x, y) = 3x + 4y$$

over this new feasible region is maximized at $x = \frac{9}{4}$ and $y = \frac{5}{4}$ with maximal value 11.75.

What this example illustrates is that for some perturbations of a feasible region's defining inequalities some perturbations of optimal corner points may continue to yield optimal corner points of the new programming problem. The same considerations may also apply for small perturbations of the objective function (in addition to small perturbations of the feasible region).

However, such considerations need not always hold true and indeed larger fluctuations of the optimal value of the objective function may also arise. As mentioned, we revisit such considerations from the view point of the Simplex Method in Example 5.9.

3.3 Canonical Form Linear Programming Problems

In studying *linear programming problems*, it is helpful to consider those that are in a certain kind of *canonical form*. (Some texts use the term *standard form* in place of what we call canonical form here.) We define what we mean by this in what follows.

In our formulation of canonical maximization and minimization linear programming problems, respectively, we allow for *affine linear* objective functions as opposed to the more restrictive setting of *linear* objective functions. Of course, we note that linear objective functions are simply those affine linear objective functions which have constant term equal to zero.

In more precise terms, an *affine linear objective function*, has the form

$$f(x_1, \ldots, x_n) = c_1x_1 + \cdots + c_nx_n - d$$

with the constant term d allowed to be nonzero. On the other hand, a *linear objective function*, has the form

$$f(x_1, \dots, x_n) = c_1 x_1 + \cdots + c_n x_n.$$

As is apparent in our approach to the Simplex Method, see Chapter 5, which is based on the theory of Tucker Tableau, it turns out that canonical linear programming problems with affine linear objective functions, as opposed to linear objective functions, play a key role.

To begin with, by a *canonical maximization linear programming problem*, sometimes called *standard form linear programming problem*, we mean a maximization linear programming problem of the following form.

- Maximize the affine linear objective function

$$f(x_1, \dots, x_n) = c_1 x_1 + \cdots + c_n x_n - d$$

- Subject to the constraints

$$\begin{array}{ccccc} a_{11}x_1 & + \dots + & a_{1n}x_n & \leq & b_1 \\ \vdots & \vdots & \vdots & & \vdots \\ a_{m1}x_1 & + \dots + & a_{mn}x_n & \leq & b_m \end{array}$$

  and

$$x_1, \dots, x_n \geq 0.$$

While not all maximization linear programming problems are canonical maximization linear programming problems, a first step in studying such, more general, maximization linear programming problems is to reduce the study of their optimal solution(s) to that of a suitable canonical maximization linear programming problem.

We may express such canonical maximization linear programming problems more succinctly in matrix form.

- Maximize the affine linear objective function

$$f(x_1, \dots, x_n) = c_1 x_1 + \cdots + c_n x_n - d = \mathbf{c}\mathbf{x}^T - \mathbf{d}$$

- Subject to the constraints

$$A\mathbf{x} \leq \mathbf{b} \text{ and } \mathbf{x} \geq \mathbf{0}.$$

  Here

$$A = \begin{bmatrix} a_{11} & \dots & a_{1n} \\ \vdots & \vdots & \vdots \\ a_{m1} & \dots & a_{mn} \end{bmatrix} \in \mathbb{R}^{m \times n}, \mathbf{b} = \begin{bmatrix} b_1 \\ \vdots \\ b_m \end{bmatrix} \in \mathbb{R}^{m \times 1},$$

$$\mathbf{c} = \begin{bmatrix} c_1 \\ \vdots \\ c_n \end{bmatrix} \in \mathbb{R}^{n\times 1} \text{ and } \mathbf{d} = [d] \in \mathbb{R}^{1\times 1};$$

whereas

$$\mathbf{x} = \begin{bmatrix} x_1 \\ \vdots \\ x_n \end{bmatrix} \in \mathbb{R}^{n\times 1}$$

is an unknown nonnegative column vector. In fact, here and elsewhere we mostly avoid the use of matrix notation.

Remark 3.4. To emphasize the difference between affine linear compared to linear objective functions, affine linear objective functions, as described above, will have the form

$$f(x_1, \ldots, x_n) = c_1x_1 + \cdots + c_nx_n - d$$

with the constant term d allowed to be nonzero. On the other hand, linear objective functions, as discussed above, will have the form

$$f(x_1, \ldots, x_n) = c_1x_1 + \cdots + c_nx_n.$$

Thus, linear objective functions are those affine linear objective functions which have constant term equal to zero.

Similar to our formulation of canonical maximization linear programming problems, we formulate *canonical minimization linear programming problems* in the following way.

- Minimize the affine linear objective function

$$g(y_1, \ldots, y_m) = b_1y_1 + \cdots + b_my_m - d$$

- Subject to the constraints

$$\begin{array}{ccccccc} a_{11}y_1 & + & \ldots & + & a_{m1}y_m & \geq & c_1 \\ \vdots & & \vdots & & \vdots & & \vdots \\ a_{1n}y_1 & + & \ldots & + & a_{mn}y_n & \geq & c_n \end{array}$$

and

$$y_1, \ldots, y_n \geq 0.$$

Similarly to the case of canonical maximization linear programming problems, at times it is helpful to express such canonical minimization linear programming problems in matrix terms. This is achieved in the following way.

- Minimize the affine linear objective function

$$g(y_1, \ldots, y_m) = b_1 y_1 + \cdots + b_m y_m - d = \mathbf{y}^T \mathbf{b} - \mathbf{d}$$

- Subject to the constraints

$$A^T \mathbf{y} \geq \mathbf{c} \text{ and } \mathbf{y} \geq \mathbf{0}.$$

Here

$$A = \begin{bmatrix} a_{11} & \ldots & a_{1n} \\ \vdots & \vdots & \vdots \\ a_{m1} & \ldots & a_{mn} \end{bmatrix} \in \mathbb{R}^{m \times n}, \mathbf{b} = \begin{bmatrix} b_1 \\ \vdots \\ b_m \end{bmatrix} \in \mathbb{R}^{m \times 1},$$

$$\mathbf{c} = \begin{bmatrix} c_1 \\ \vdots \\ c_n \end{bmatrix} \in \mathbb{R}^{n \times 1} \text{ and } \mathbf{d} = [d] \in \mathbb{R}^{1 \times 1};$$

whereas

$$\mathbf{y} = \begin{bmatrix} y_1 \\ \vdots \\ y_m \end{bmatrix} \in \mathbb{R}^{m \times 1}$$

is an unknown nonnegative column vector. Again, here and elsewhere we mostly avoid the use of matrix notation.

Remark 3.5. Our notational conventions for expressing canonical maximization/canonical minimization linear programming problems, for instance using decision variables $x_1, \ldots, x_n$ for maximization linear programming problems compared to $y_1, \ldots, y_m$ for minimization linear programming problems, arises from the theory of dual pairs of maximization/minimization linear programming problems. This concept of dual canonical form linear programming problem is discussed in detail in Section 4.3.

Example 3.6. The resource allocation maximization linear programming model (Section 2.1) gives rise to canonical maximization linear programming problems. Indeed, the maximization linear programming problem that is formulated there takes the following form.

- Maximize the linear objective function

$$Z = f(x_1, \ldots, x_n) = c_1 x_1 + \cdots + c_n x_n$$

- Subject to the constraints

$$\begin{array}{ccccc} a_{11}x_1 & + \ldots + & a_{1n}x_n & \leq & b_1 \\ \vdots & \vdots & \vdots & & \vdots \\ a_{m1}x_1 & + \ldots + & a_{mn}x_n & \leq & b_m \end{array}$$

and

$$x_1, \ldots, x_n \geq 0.$$

Example 3.7. The model for nutrition diet problems, with no restriction on the number of servings, (Section 2.2) gives rise to canonical minimization linear programming problems. Indeed, the minimization linear programming problem that is formulated there takes the following form.

- Minimize the objective function

$$Z = g(y_1, \ldots, y_m) = d_1 y_1 + \cdots + d_m y_m$$

- Subject to the constraints

$$\begin{aligned} e_1 y_1 + \ldots + e_m y_m &\geq E \\ p_1 y_1 + \ldots + p_m y_m &\geq P \\ c_1 y_1 + \ldots + c_m y_m &\geq C \end{aligned}$$

and

$$y_1, \ldots, y_m \geq 0.$$

Example 3.8. Both of the models for mixture blending with production levels and selling strategies (Section 2.3) and the model for agricultural planning (Section 2.4) are not canonical maximization/minimization programming problems. Nonetheless, following the techniques that are described in Section 3.5, optimal solutions can be found upon first having given suitable reformulations of such problems which reduce their solutions to those of suitably formulated canonical form linear programming problems.

Example 3.9. Multiplication by -1 transforms each given canonical maximization linear programming problem

- Maximize the affine linear objective function

$$f(x_1, \ldots, x_n) = c_1 x_1 + \cdots + c_n x_n - d$$

- Subject to the constraints

$$\begin{array}{ccccccc} a_{11}x_1 & + & \ldots & + & a_{1n}x_n & \leq & b_1 \\ \vdots & & \vdots & & \vdots & & \vdots \\ a_{m1}x_1 & + & \ldots & + & a_{mn}x_n & \leq & b_m \end{array}$$

and

$$x_1, \dots, x_n \geq 0$$

into a canonical minimization linear programming problem

- Minimize the affine linear objective function

$$-f(x_1, \dots, x_n) = -c_1x_1 - \cdots - c_nx_n + d$$

- Subject to the constraints

$$\begin{array}{cccc} -a_{11}x_1 & - \dots - & a_{1n}x_n & \geq -b_1 \\ \vdots & \vdots & \vdots & \vdots \\ -a_{m1}x_1 & - \dots - & a_{mn}x_n & \geq -b_m \end{array}$$

and

$$x_1, \dots, x_n \geq 0.$$

And vice versa.

In studying such canonical form maximization, respectively, minimization, linear programming problems, the set of points

$$\mathbf{x} = (x_1, \dots, x_n) \in \mathbb{R}^n$$

that satisfy each of the given $m + n$ constraints

$$\begin{array}{cccc} a_{11}x_1 & + \dots + & a_{1n}x_n & \leq b_1 \\ \vdots & \vdots & \vdots & \vdots \\ a_{m1}x_1 & + \dots + & a_{mn}x_n & \leq b_m \end{array}$$

and

$$x_1, \dots, x_n \geq 0$$

respectively, the set of points

$$\mathbf{y} = (y_1, \dots, y_m) \in \mathbb{R}^m$$

that satisfy each of the given $m + n$ constraints

$$\begin{array}{cccc} a_{11}y_1 & + \dots + & a_{m1}y_m & \geq c_1 \\ \vdots & \vdots & \vdots & \vdots \\ a_{1n}y_1 & + \dots + & a_{mn}y_m & \geq c_n \end{array}$$

and

$$y_1, \ldots, y_m \geq 0$$

is called the *constraint set* or, equivalently, the *feasible region*.

Each feasible solution of a canonical maximization (respectively, minimization) linear programming problem which maximizes (respectively, minimizes) the objective function is said to be an *optimal solution*. At times we find the need to fail to distinguish between the concepts of canonical maximization and canonical minimization linear programming problems. When these needs arise, we simply refer to such problems as *canonical linear programming problems*.

In doing so, we understand that a canonical linear programming problem is either a maximization problem or a minimization problem.

3.4 Slack Formulation of Canonical Linear Programming Problems

The first step to solve a given canonical maximization linear programming problem

- Maximize the affine linear objective function

$$f(x_1, \ldots, x_n) = c_1x_1 + \cdots + c_nx_n - d$$

- Subject to the constraints

$$\begin{array}{ccccc} a_{11}x_1 & + \ldots + & a_{1n}x_n & \leq & b_1 \\ \vdots & \vdots & \vdots & & \vdots \\ a_{m1}x_1 & + \ldots + & a_{mn}x_n & \leq & b_m \end{array}$$

 and

$$x_1, \ldots, x_n \geq 0$$

is to introduce auxiliary *slack variables* $x_{n+1}, \ldots, x_{m+n}$ and then write it in *slack form*

- Maximize the affine linear objective function

$$f(x_1, \ldots, x_n) = c_1x_1 + \cdots + c_nx_n - d$$

- Subject to the constraints

$$\begin{array}{ccccccc} a_{11}x_1 & + \ldots + & a_{1n}x_n & + & x_{n+1} & = & b_1 \\ \vdots & \vdots & \vdots & & \vdots & & \vdots \\ a_{m1}x_1 & + \ldots + & a_{mn}x_n & + & x_{m+n} & = & b_m \end{array}$$

and

$$x_1, \ldots, x_{m+n} \geq 0.$$

Similarly, the first step to solving a given canonical minimization linear programming problem

- Minimize the affine linear objective function

$$g(y_1, \ldots, y_m) = b_1 y_1 + \cdots + b_m y_m - d$$

- Subject to the constraints

$$\begin{array}{ccccc} a_{11}y_1 & + \ldots + & a_{m1}y_m & \geq & c_1 \\ \vdots & \vdots & \vdots & & \vdots \\ a_{1n}y_1 & + \ldots + & a_{mn}y_m & \geq & c_n \end{array}$$

and

$$y_1, \ldots, y_m \geq 0$$

is to introduce auxiliary *slack variables* $y_{m+1}, \ldots, y_{m+n}$ and then write it in *slack form*

- Minimize the affine linear objective function

$$g(y_1, \ldots, y_m) = b_1 y_1 + \cdots + b_m y_m - d$$

- Subject to the constraints

$$\begin{array}{ccccc} a_{11}y_1 & + \ldots + & a_{m1}y_m & = & c_1 + y_{m+1} \\ \vdots & \vdots & \vdots & & \vdots \\ a_{1n}y_1 & + \ldots + & a_{mn}y_m & = & c_n + y_{m+n} \end{array}$$

and

$$y_1, \ldots, y_{m+n} \geq 0.$$

The usefulness for reformulating such canonical form maximization and/or minimization linear programming problems via their associated slack forms are apparent upon discussing the concept of Tucker Tableau as a tool for the Simplex Method. We postpone further discussion about such considerations for the present time being. They are developed in detail in Chapters 4 and 5.

Here we do note that a key point for introducing slack variables is that they allow for the study of systems of linear inequalities to be reduced to the study of linear equalities. Such slack reformulations of linear programming problems facilitate with finding extreme points of the feasible region. In fact, as is a consequence of the Simplex Method, performing a suitable sequence of Pivot Transformations to such slack form linear programming problems, allows for finding extreme points which yield optimal values (should any exist) for the objective function.

3.5 More General Linear Programming Problems

In Section 3.3, we introduced the concept of canonical maximization/minimization linear programming problems. As we will see in Chapter 5, such linear programming problems have a special structure which facilitates with finding an optimal solution via the Simplex Method. Here, we consider more general linear programming problems and indicate some techniques to reduce their solution to those of suitably formulated canonical slack form linear programming problems.

Example 3.10 (Nonnegativity conditions). Consider the case of a maximization linear programming problem of the form

- Maximize the affine linear objective function

$$f(x_1, \ldots, x_n) = c_1x_1 + \cdots + c_nx_n - d$$

- Subject to the constraints

$$\begin{array}{ccccc} a_{11}x_1 & + \ldots + & a_{1n}x_n & \le & b_1 \\ \vdots & \vdots & \vdots & & \vdots \\ a_{m1}x_1 & + \ldots + & a_{mn}x_n & \le & b_m \end{array}$$

  and

$$x_1, \ldots, x_\ell \ge 0 \text{ for some } \ell,\ 1 \le \ell < n.$$

Especially, observe that there are no nonnegativity constraints on $x_{\ell+1}, \ldots, x_n$.

Now, for each $j = \ell + 1, \ldots, n$ set

$$x_j := t_j - w_j$$

where

$$t_j \ge 0 \text{ and } w_j \ge 0.$$

Then each of the ith constraints of the given maximization linear programming problem can be re-expressed in the form

$$a_{i1}x_1 + \cdots + a_{i\ell}x_\ell + a_{i\ell+1}t_{\ell+1} - a_{i\ell+1}w_{\ell+1} + \cdots + a_{in}t_n - a_{in}w_n \le b_i$$

for $i = 1, \ldots, m$.

In such a way, the given maximization linear programming problem has been transformed into a canonical form maximization linear programming problem.

This has been achieved by replacing each variable that is not required to be non-negative by the difference of two variables that is required to be nonnegative. Similar considerations apply to linear programming problems in which not all decision variables are required to be nonnegative.

Example 3.11 (Slack variables)**.** Suppose that a linear maximization/minimization linear programming problem contains linear constraints of the form

$$a_{i1}x_1 + \cdots + a_{in}x_i \leq b_i.$$

Each such constraint can be converted into an equality by adding a new nonnegative variable to the left-hand side of the inequality. Specifically, each such constraint can be replaced by

$$a_{i1}x_1 + \cdots + a_{in}x_i + t_i = b_i$$

with $t_i \geq 0$. The variable t_i is referred to as a *slack variable.*

Such slack variables represent the *waste* that is modelled by the given $\leq$ constraints. Note that such a technique has been employed when deriving the slack form of a canonical maximization linear programming problem.

Example 3.12 (Surplus variables)**.** Consider the case that a given maximization/minimization linear programming problem contains linear constraints of the form

$$a_{i1}x_1 + \cdots + a_{in}x_i \geq b_i.$$

Each such constraint can be converted into an equality by subtracting a new nonnegative variable from the left-hand side of the inequality. Specifically, each such constraint can be replaced by

$$a_{i1}x_1 + \cdots + a_{in}x_i = b_i + s_i.$$

with $s_i \geq 0$. The extra variables that have been introduced are sometimes called *surplus variables.*

The idea is that surplus variables represent the *excess* input that is modelled by the constraint. Note that such a technique has been employed when deriving the slack form of a canonical minimization linear programming problem.

Example 3.13 (General linear constraints)**.** Recall, that in a linear programming problem, the linear constraints are assumed to be of the form

$$\sum_{j=1}^{n} a_{ij}x_j \ ? \ b_i \text{ for } i = 1, \ldots, n$$

where ? is any of the relations $\leq, \geq$ or $=$. There are the additional nonnegativity constraints $x_1, \ldots, x_n \geq 0$. Finally, by multiplying a given constraint by -1 and replacing $\leq$ by $\geq$ and/or $\geq$ by $\leq$, if required, there is no loss in generality by assuming that $b_i \geq 0$ for $i = 1, \ldots, n$.

Similarly, by multiplying a given constraint by -1 and replacing $\leq$ by $\geq$ and/or $\geq$ by $\leq$ if required there is no loss generality by assuming that either all constraints are either of the form $\leq$ and/or $=$ or $\geq$ and/or $=$. Such techniques can be combined, if necessary, with those that are described in Example 3.10, to re-express linear programming problems that involve a combination of $\leq$ and $\geq$ into canonical form maximization/minimization linear programming problems.

Example 3.14 (Generating initial feasible solutions)**.** Continuing with the setting that is described in Example 3.13, after all linear constraints with nonnegative right-hand sides have been transformed into equalities by introducing slack and surplus variables, where required, add a new variable called an *artificial variable* to the left-hand side of each constraint equation that does not contain a slack variable.

Each constraint equation will then contain either one slack variable or one artificial variable. This set-up facilitates the process of generating initial feasible solutions. For example, a nonnegative initial solution to this new set of constraints is obtained by setting each slack variable and each artificial variable equal to the right-hand side of the equation in which it appears and then setting all other variables—including the surplus variables equal to zero.

Example 3.15 (Penalty Costs)**.** Continuing with the setting that is described in Example 3.13, and after introducing slack, surplus and artificial variables, note that the slack and surplus variables in a given linear programming problem have no actual effect on the nature of the constraints and/or the given objective function. For example, such variables are incorporated into the objective function (implicitly) with zero coefficients.

On the other hand, note that introducing artificial variables do change the nature of the constraints. Indeed, since they are added to only one side of an equality, the new system is equivalent to the old system if and only if the artificial variables are zero. In order to account for this, and to guarantee assignments in the optimal solution, artificial variables are incorporated into the objective function with very large positive coefficients in a minimization problem and very large negative coefficients in a maximization problem.

These coefficients, denoted either by M or $-M$, and where M is understood to be a very large positive number, represent the *severe penalty* that is incurred in the making an assignment to the artificial variables.

Example 3.16 (Achieving canonical form linear programming problems)**.** Continuing with the setting that is described in Example 3.13, observe that after introducing slack variables, surplus variables, artificial variables and penalty costs (if required) a given linear programming problem, with linear objective function, can be converted into a *canonical form* maximization/linear programming problem. Thus as a consequence, in the development of algorithmic tools for solving linear programming problems, in essence, there is no loss in generality by placing emphasis on canonical form maximization/minimization linear programming problems.

Example 3.17 (Integer and binary linear programming problems)**.** In real-world applications, it is oftentimes required to impose the conditions that the decision variables are integers or even that they are binary integers 0 or 1. Respectively, such linear programming problems are called *integer programming problems* (when all decision variables are required to be integers) and *binary integer programming problems* (when all decision variables are required to be either 0 or 1).

In general, the solving of integer programming problems is quite complicated and requires specialized techniques. On the other hand, many important classes of integer programming problems that are of interest for our purposes here, including the transportation, assignment and network flow problems (see Sections 7.2, 7.4 and 8.3), turn out to possess a sort of *integer solutions property*.

In particular, such integer programming problems have the property that all *basic variables* in each *basic feasible solution* (including the optimal one) have integer values. Among other things, this allows such integer programming problems (which have the integer solutions property) to be solved via the Simplex Method which applies to canonical form linear programming problems (and with no actual requirement that the decision variables have integer values). The concept of basic and non-basic variables together with the concept of basic feasible solution is discussed, within the context of Tucker Tableau, in Chapters 4 and 5.

Example 3.18 (Branch and bound techniques for integer programming problems)**.** In this example, we illustrate how *branch and bound* techniques can be used to solve integer linear programming problems.

To this end, consider the following integer programming problem.

- Maximize the objective function

$$f(x_1, x_2) = 4x_1 + 6x_2$$

- Subject to the constraints

$$\begin{aligned} x_1 + x_2 &\leq 4 \\ 3x_1 + 5x_2 &\leq 15 \end{aligned}$$

  and

$$x_1, x_2 \geq 0 \text{ and } x_1, x_2 \text{ are integers}$$

together with the following description of the feasible region of the associated *relaxation linear programming problem*, that is the linear programming problem that is obtained by omitting the requirement that the decision variables are required to be integers. The feasible region is described in Figure 3.5.

It has corner points

$$(0, 0), (0, 3), \left(\frac{5}{2}, \frac{3}{2}\right) \text{ and } (4, 0).$$

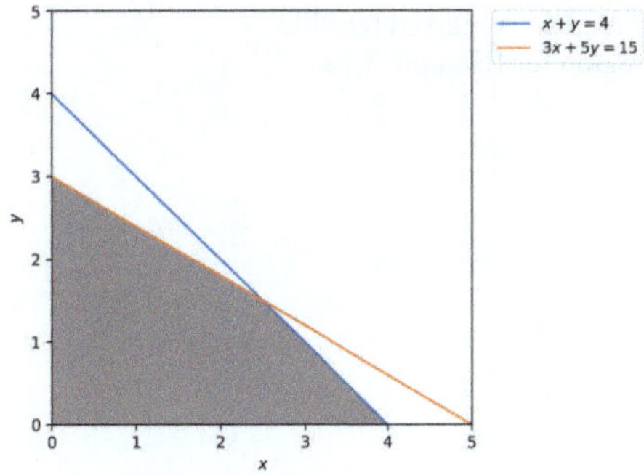

Fig. 3.5 The feasible region for Example 3.18

In particular, it follows that the relaxation of the given integer programming problem has optimal value 19 and that this is achieved when $x_1 = \frac{5}{2}$ and $x_2 = \frac{3}{2}$. In terms of the given integer programming problem at hand, it follows that its optimal value is at most equal to 19.

To solve the given integer programming problem, the idea is to observe that its optimal solutions have either $x_1 \geq 3$ or $x_1 \leq 2$. We thus consider separately the following two integer programming problems.

- Maximize the objective function

$$f(x_1, x_2) = 4x_1 + 6x_2$$

- Subject to the constraints

$$\begin{aligned} x_1 + x_2 &\leq 4 \\ 3x_1 + 5x_2 &\leq 15 \\ -x_1 \quad &\leq -3 \end{aligned}$$

  and

$$x_1, x_2 \geq 0 \text{ and } x_1, x_2 \text{ are integers.}$$

Respectively,

- Maximize the objective function

$$f(x_1, x_2) = 4x_1 + 6x_2$$

- Subject to the constraints

$$\begin{aligned} x_1 + x_2 &\leq 4 \\ 3x_1 + 5x_2 &\leq 15 \\ x_1 \quad &\leq 2 \end{aligned}$$

  and

$$x_1, x_2 \geq 0 \text{ and } x_1, x_2 \text{ are integers.}$$

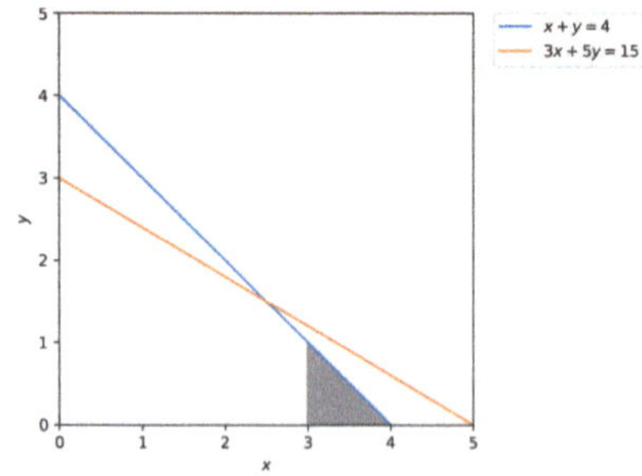

Fig. 3.6 A relaxed feasible region for Example 3.18

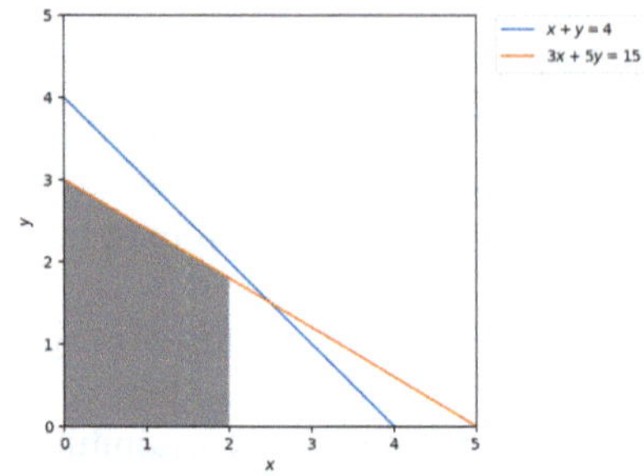

Fig. 3.7 A relaxed feasible region for Example 3.18

The relaxation of each of these two linear programming problems have respective feasible regions that are pictured in Figures 3.6 and 3.7.

The objective function $f(x_1, x_2)$ is maximized respectively at the respective corner points (3, 1) and (2, 2) and with respective maximum value 18 and 20. Since these corner points are integral, it follows that the optimal solution to the given maximization integer programming problem is 20 which is achieved when $x_1 = 2$ and $x_2 = 2$.

This same *branch and bound* methods works quite generally to solve, in a reasonably efficient way, integer programming problems in general.

Example 3.19 (Either-or constraints)**.** Given a mathematical programming problem which has a constraint of the form

$$\text{either } f(x_1, \ldots, x_n) \leq 0 \text{ or } g(x_1, \ldots, x_n) \leq 0$$

upon choosing a sufficiently large integer M and introducing an auxiliary binary variable y, the given *either-or constraint* can be replaced by a pair of constraints which have shape

$$f(x_1, \ldots, x_n) \leq My \text{ and } g(x_1, \ldots, x_n) \leq M(1 - y) \text{ where } y = 0 \text{ or } 1.$$

In choosing such sufficiently large values of M, the idea is to choose M large enough so that

$$f(x_1, \ldots, x_n) \leq M \text{ and } g(x_1, \ldots, x_n) \leq M$$

for all feasible values of $x_1, \ldots, x_n$.

Example 3.20 (If-then constraints). Given a mathematical programming problem which has a constraint of the form

$$\text{if } f(x_1, \ldots, x_n) > 0 \text{ then } g(x_1, \ldots, x_n) \geq 0$$

then upon choosing a sufficiently large integer M and introducing an auxiliary binary variable y, the given *if-then constraint* can be replaced by a pair of constraints which have shape

$$-g(x_1, \ldots, x_n) \leq My \text{ and } f(x_1, \ldots, x_n) \leq M(1-y) \text{ where } y = 0 \text{ or } 1.$$

In choosing a sufficiently large value of M, the idea is to ensure that

$$f(x_1, \ldots, x_n) \leq M \text{ and } -g(x_1, \ldots, x_n) \leq M$$

for all feasible values of $x_1, \ldots, x_n$.

3.6 Convex Geometric Considerations

There exists a rich underlying convex geometry that surrounds linear programming problems. We indicate some of the most basic considerations here. They should be familiar from multivariable calculus.

To begin with, if

$$\mathbf{x} = (x_1, \ldots, x_n) \text{ and } \mathbf{y} = (y_1, \ldots, y_n) \in \mathbb{R}^n$$

then the *line segment* that is between them is described as

$$\{\lambda\mathbf{x} + (1-\lambda)\mathbf{y} : 0 \leq \lambda \leq 1\}.$$

Further, a subset

$$S \subseteq \mathbb{R}^n$$

is *convex* if

$$\lambda\mathbf{x} + (1-\lambda)\mathbf{y} \in S \text{ for all } 0 \leq \lambda \leq 1 \text{and all } \mathbf{x}, \mathbf{y} \in S.$$

Example 3.21. In Figure 3.8 below, the leftmost region is convex, whereas the rightmost region is not convex.

Indeed, the leftmost region contains the line segment between each pairs of its points. Hence, it is convex. On the other hand, the rightmost region does not contain the line segment between some pairs of its points. Hence, it is not convex.

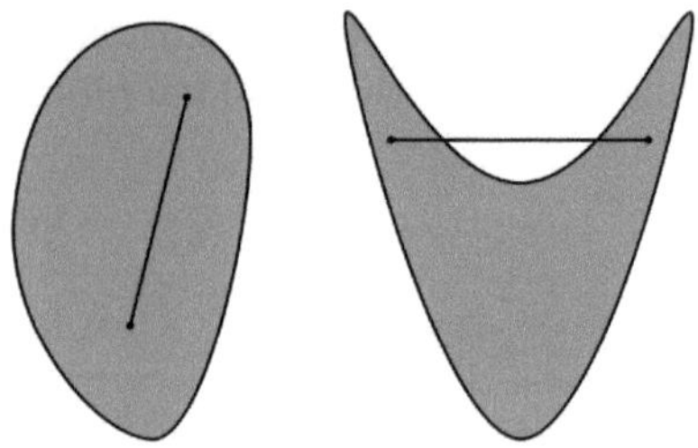

Fig. 3.8 A convex and non-convex region for Example 3.21

By an *affine hyperplane* in $\mathbb{R}^n$, is meant the collection of points

$$\mathbf{x} = (x_1, \dots, x_n) \in \mathbb{R}^n$$

that satisfy an equation of the form

$$a_1 x_1 + \cdots + a_n x_n = b.$$

Finally, a *closed half space* is defined to be the collection of points

$$\mathbf{x} = (x_1, \dots, x_n) \in \mathbb{R}^n$$

that satisfy an inequality of shape

$$a_1 x_1 + \cdots + a_n x_n \leq b.$$

The most basic structure of the feasible region of a given canonical programming problem is made precise in the following way.

Theorem 3.22. *Let $S \subseteq \mathbb{R}^n$ be the constraint set of a canonical linear programming problem. Then S is a polyhedral convex set. In particular, it is a finite intersection of closed half spaces.*

In the statement of Theorem 3.22, *polyhedral* reflects the fact that the boundaries are *affine hyperplanes*.

Example 3.23. The feasible region described in Figure 3.5, is the intersection of the closed half spaces

$$x + y \leq 4,\ 3x + 5y \leq 15,\ x \geq 0 \text{ and } y \geq 0.$$

The boundaries of these respective closed half spaces are the affine hyperplanes

$$x + y = 4,\ 3x + 5y = 15,\ x = 0 \text{ and } y = 0.$$

Proof of Theorem 3.22. As mentioned, in the statement of Theorem 3.22, the assertion that the feasible region S is *polyhedral* reflects the fact that the boundaries are affine hyperplanes. That S is a finite intersection of closed half spaces is evident as the feasible region is defined as the solution set of a finite number of linear inequalities. That S is convex also is clear since a finite intersection of closed half spaces is convex. □

We want to study the *extreme points* of a constraint set and also the extent to which a given constraint set is unbounded. It turns out that the extreme points interact fruitfully with the solutions of linear programming problems.

To formulate precise statements, we define the *norm* of

$$\mathbf{x} = (x_1, \ldots, x_n) \in \mathbb{R}^n$$

to be

$$||\mathbf{x}|| := \sqrt{x_1^2 + \cdots + x_n^2}.$$

Further, given a nonnegative real number $r \geq 0$, the *closed ball* with radius r and centred at the origin

$$\mathbf{0} \in \mathbb{R}^n$$

is defined to be

$$B(r; \mathbf{0}) := \left\{\mathbf{x} = (x_1, \ldots, x_n) \in \mathbb{R}^n : ||x|| \leq r\right\}.$$

Finally, a subset

$$S \subseteq \mathbb{R}^n$$

is said to be *bounded* if there exists a nonnegative real number $r \geq 0$ which is such that each $x \in S$ is contained in some closed ball $B(r; \mathbf{0})$; whereas a subset

$$S \subseteq \mathbb{R}^n$$

that is not bounded is said to be *unbounded*.

Now, let

$$S \subseteq \mathbb{R}^n$$

be a convex set. Let $\mathbf{w} \in S$. Then $\mathbf{w}$ is an *extreme point* if there does not exist $\mathbf{x}, \mathbf{y} \in S$ and $0 < \lambda < 1$ which are such that

$$\mathbf{w} = \lambda\mathbf{x} + (1 - \lambda)\mathbf{y}.$$

For the case that S is the *constraint set* (that is *feasible region*) of a linear programming problem, then we oftentimes use the term *corner point* in place of extreme point.

3.7 Theoretical Justification of the Main Geometric Method

The main geometric methods for solving linear programming problems ultimately are extensions to the following theorem.

Theorem 3.24. *The following assertions hold true.*

(i) If the constraint set S of a canonical maximization, respectively, canonical minimization problem is bounded, then the maximum, respectively, minimum, value of the objective function is attained at an extreme point of S.

(ii) If the constraint set S of a canonical maximization linear programming problem is unbounded and if the objective function is bounded from above on S, then the maximum value of the objective function is attained at an extreme point of S.

(iii) If the constraint set S of a canonical minimization linear programming problem is unbounded and if the objective function is bounded from below on S, then the minimum value of the objective function is attained at an extreme point of S.

In the statement of Theorem 3.24, the first item is really a special case of the second and third items since if the constraint set $S \subseteq \mathbb{R}^n$ is bounded, then it is a closed and bounded subset of $\mathbb{R}^n$ and hence is compact; the objective function, which is continuous on S, will thus be bounded from below and above.

A consequence of Theorem 3.24 is essentially that canonical linear programming problems can be solved by evaluating the objective function at the extreme points of the constraint set. To put this exhaustive search geometric method into perspective, it is helpful to keep in mind that the constraint set will contain at most

$$\binom{m+n}{n} = \frac{(m+n)!}{m!n!}$$

extreme points.

Let us now turn to a more detailed discussion of the proof of Theorem 3.24. There are a few steps. First, the key point is to verify that optimal solutions to the given canonical form linear programming problem can be achieved at some extreme point of the feasible region. For this purpose, the key technical observation is expressed as Lemma 3.25 below.

With Lemma 3.25 in hand, the conclusions of assertions (i), (ii) and (iii) in Theorem 3.24 are readily deduced. For example, via the concept of negative transpose to convert a given canonical form minimization linear programming problem into a canonical form maximization linear programming problem and vice versa it suffices to prove assertions (i) and (ii) for the case of canonical form maximization linear programming problems.

The conclusions of Theorem 3.24 are also naturally explained via the concept of duality for linear programming problems (see Section 5.6).

For this purpose, the key technical observation is expressed as Lemma 3.25 below.

Lemma 3.25. *Consider a canonical maximization linear programming problem*

- *Maximize the affine linear objective function*

$$f(x_1, \ldots, x_n) = c_1x_1 + \cdots + c_nx_n - d$$

- *Subject to the constraints*

$$\begin{array}{cccc} a_{11}x_1 & + \ldots + & a_{1n}x_n & \leq b_1 \\ \vdots & \vdots & \vdots & \vdots \\ a_{m1}x_1 & + \ldots + & a_{mn}x_n & \leq b_m \end{array}$$

and

$$x_1, \ldots, x_n \geq 0.$$

Assume that the objective function $f(x_1, \ldots, x_n)$ is bounded above on the feasible region S. Then an optimal value of the objective function is obtained at an extreme point of the feasible region.

Proof of Lemma 3.25. By assumption, $f(x_1, \ldots, x_n)$ is bounded above on S. Thus $f(x_1, \ldots, x_n)$ is maximized by some

$$\mathbf{x}^* = (x_1^*, \ldots, x_n^*) \in S.$$

Suppose that $\mathbf{x}^*$ is not an extreme point of S. Then

$$f(q_1, \ldots, q_n) \leq f(x_1^*, \ldots, x_n^*) \text{ for all } \mathbf{q} = (q_1, \ldots, q_n) \in S.$$

On the other hand, by convexity of the feasible region S, $\mathbf{x}^*$ can be written as a convex linear combination of extreme points $\mathbf{p}_1, \ldots, \mathbf{p}_N \in S$. Therefore

$$\mathbf{x}^* = \sum_{i=1}^{N} \lambda_i \mathbf{p}_i$$

for extreme points

$$\mathbf{p}_1 = (p_{11}, \ldots, p_{1n}), \ldots, \mathbf{p}_N = (p_{N1}, \ldots, p_{Nn}) \in S$$

and nonnegative real numbers $\lambda_i \in \mathbb{R}_{\geq 0}$, for $i = 1, \ldots, N$, which are such that

$$1 = \sum_{i=1}^{N} \lambda_i.$$

In then follows that

$$
\begin{aligned}
f(\mathbf{x}^*) = c_1 x_1^* + \cdots + c_n x_n^* - d &= c_1 \left(\sum_{i=1}^{N} \lambda_i p_{i1} \right) + \cdots + c_n \left(\sum_{i=1}^{N} \lambda_i p_{in} \right) - d \\
&\leq \max_{1 \leq k \leq N} \left(\sum_{i=1}^{N} \lambda_i \right) (c_1 p_{k1} + \cdots + c_n p_{kn}) - d \\
&= \left(\sum_{i=1}^{N} \lambda_i \right) \max_{1 \leq k \leq N} (c_1 p_{k1} + \cdots + c_n p_{kn}) - d \\
&\leq \max_{1 \leq k \leq N} f(\mathbf{p}_k).
\end{aligned}
$$

That $f(x_1, \ldots, x_n)$ is indeed maximized at an extreme point is then evident. □

Proof of Theorem 3.24. Via the operation of negative transpose to convert a given canonical form minimization linear programming problem into a canonical form maximization linear programming problem and vice versa it suffices to prove assertions (i) and (ii) for the case of canonical form maximization linear programming problems. Evidently, the conclusion of assertions (i) and (ii) applied to the case of canonical form maximization linear programming problems are mutually exclusive. The conclusion that an optimal value is achieved at an extreme point in either case follows from Lemma 3.25. □

Exercises

3.1 Consider the linear programming problem

- Minimize the linear objective function

$$g(x, y, z, w) = x - 2y + 3z - 4w$$

- Subject to the constraints

$$-5x - 4y - 3z - 2w \geq -1$$

and

$$x, y, z, w \geq 0.$$

(i) Show that the constraint set is bounded.
(ii) Find all potential extreme points by considering the given 5 constraints as equations and then by solving the $\binom{5}{4}$ systems of linear equations obtained from these constraints by taking them four at a time.
(iii) Solve the given linear programming problem by applying the geometric method.

3.2 Consider the following linear programming problem

- Maximize the linear objective function

$$Z = x_1 + 2x_2$$

- Subject to the constraints

$$\begin{aligned} x_1 \quad\quad &\le 2 \\ x_2 &\le 2 \\ x_1 + x_2 &\le 3 \end{aligned}$$

and

$$x_1, x_2 \ge 0.$$

(i) Plot the feasible region and identify all corner point solutions.
(ii) For each corner point solution, identify the pair of constraint boundary equations that it satisfies.
(iii) For each corner point solution, use this pair of constraint boundary equations to solve algebraically for the values of x_1 and x_2 at the corner point.
(iv) For each pair of adjacent corner point solutions, identify the constraint boundary that they share by giving its equation.

3.3 Suppose that a certain linear programming model involving two activities has feasible region with corner point solutions

$$(0, 0), \left(0, \frac{20}{8}\right), (5, 5), (6, 4), (8, 0) \in \mathbb{R}^2.$$

Further, suppose that the aim is to maximize the total profit from the two activities. The unit profit for Activity 1 is \$1, 000 and the unit profit for Activity 2 is \$2, 000. Calculate the total profit for each corner point solution and use this information to find an optimal solution.

3.4 Consider the maximization linear programming problem

- Maximize the linear objective function

$$f(x_1, x_2) = 20x_1 + 10x_2$$

- Subject to the constraints

$$\begin{aligned} x_1 + 2x_2 &\le 40 \\ 3x_1 + 2x_2 &\le 60 \end{aligned}$$

and

$$x_1, x_2 \ge 0.$$

Determine the corresponding slack maximization linear programming problem.

3.5 Consider the minimization linear programming problem

- Minimize the linear objective function

$$g(y_1, y_2) = 10y_1 + 6y_2 + 8y_2$$

- Subject to the constraints

$$\begin{aligned} y_1 + y_2 + 2y_3 &\geq 2 \\ 5y_1 + 3y_2 + 2y_3 &\geq 1 \end{aligned}$$

and

$$y_1, y_2, y_3 \geq 0.$$

Determine the corresponding slack minimization linear programming problem.

3.6 Re-express the maximization linear programming problem

- Maximize the linear objective function

$$Z = -x_1 - x_2 - x_3$$

- Subject to the constraints

$$\begin{aligned} x_1 + 3x_2 \quad &\leq 10 \\ x_1 + x_2 + x_3 &\geq 6 \\ x_1 \quad &\geq 2 \end{aligned}$$

and

$$x_1, x_2, x_3 \geq 0$$

as a suitable canonical form maximization linear programming problem.

3.7 Re-express the minimization linear programming problem

- Minimize the linear objective function

$$Z = x_1 + 2x_2 + 3x_3$$

- Subject to the constraints

$$\begin{aligned} x_1 - 2x_2 + x_3 &\leq 1 \\ 3x_1 - 4x_2 + 6x_3 &\leq 8 \\ 2x_1 - 4x_2 + 2x_3 &\leq 2 \end{aligned}$$

and

$$x_1, x_2, x_3 \geq 0$$

as a suitable canonical form minimization linear programming problem.

3.8 Re-express the maximization linear programming problem

- Maximize the linear objective function

$$Z = 2x_1 + 3x_2 + 4x_3$$

- Subject to the constraints

$$\begin{aligned} x_1 + x_2 + x_3 &\leq 1 \\ x_1 + x_2 + 2x_3 &= 2 \\ 3x_1 + 2x_2 + x_3 &\geq 4 \end{aligned}$$

and

$$x_1, x_2, x_3 \geq 0$$

as a suitable canonical form maximization linear programming problem.

3.9 Re-express the minimization linear programming problem

- Minimize the linear objective function

$$Z = x_1 + 2x_2 + 3x_3$$

- Subject to the constraints

$$\begin{aligned} 3x_1 \quad\quad + 4x_3 &\leq 5 \\ 5x_1 + x_2 + 6x_3 &= 7 \\ 8x_1 \quad\quad + 9x_3 &\leq 2 \end{aligned}$$

and

$$x_1, x_2, x_3 \geq 0$$

as a suitable canonical form minimization linear programming problem.

3.10 An auto plant is able to produce compact, mid- and full size cars. Management would like to determine production levels that will maximize profit. Suppose that the required resources and profits are summarized in Table 3.3. Suppose that the plant is allocated 60, 000 kg of steel and 60, 000 hrs of labour for the upcoming fiscal year. Further, suppose that in order for production of a particular type of car to be economically viable at least 1, 000 cars of that type must be produced. Formulate and solve a suitable integer programming problem that will maximize the profit.

Table 3.3 Data table for Exercise 3.10

	Compact	Midsize	Fullsize
Steel required	840 Kg	900 Kg	1000 Kg
Labour required	31 hrs	26 hrs	37 hrs
Profit ($)	$2,000	$3,000	$4,000

Table 3.4 Payment table for Exercise 3.11

Currency	Avg. daily payment
Currency 1	\$700
Currency 2	\$500
Currency 3	\$600
Currency 4	\$400

Table 3.5 Processing time table for Exercise 3.11

	Currency 1	Currency 2	Currency 3	Currency 4
Location 1	2	5	5	5
Location 2	3	1	3	5
Location 3	5	3	1	3
Location 4	5	5	4	2

3.11 An international non-profit company receives donations in four international currencies. Suppose that the average daily value of payment by currency is summarized in Table 3.4.

Because the non-profit is able to make 2.5% annual interest by investing in its domestic currency, it would like to convert its payments to the domestic currency as soon as possible. To facilitate with this, the non-profit company would like to set up bank accounts in its domestic currency at four international locations.

Suppose that the total average processing time (days from time payment is sent, cleared and converted into the desired currency) is given in Table 3.5.

Suppose that the annual cost of running the international based bank accounts and processing fees for converting the foreign currencies into the desired currency is \$500. Formulate and solve a suitable integer programming problem that will minimize the total costs due to lost interest and bank account operations.

Chapter 4
Introduction to Tucker Tableau and Duality Theory for Linear Programming

Overview

- Introduction to Tucker Tableau.
- Introduction to duality theory for linear programming.

4.1 A First Example

Example 4.1 below illustrates the main ideas behind the Tucker Tableau approach to the Simplex Method. The undefined terms *Tucker Tableau* and *pivot* applied to a given Tucker Tableau that arise within this example are defined in Sections 4.2 and 5.3.1. The idea of Tucker Tableau is simply to express the data needed to define a given canonical maximization linear programming problem in a way that is helpful for the purposes of the Simplex Method.

Example 4.1. Consider the canonical maximization linear programming problem

- Maximize the linear objective function

$$P(x_1, x_2) = 200x_1 + 150x_2$$

- Subject to the constraints

$$\begin{aligned} x_1 + 2x_2 &\leq 20 \\ 2x_1 + 2x_2 &\leq 30 \\ 2x_1 + x_2 &\leq 25 \end{aligned}$$

  and

$$x_1, x_2 \geq 0.$$

N. Grieve, *An Introduction to Mathematical Programming and Network Science*, Springer Undergraduate Texts in Mathematics and Technology,
https://doi.org/10.1007/978-3-032-13330-4_4

The corresponding slack formulation is

- Maximize the linear objective function

$$P(x_1, x_2) = 200x_1 + 150x_2$$

- Subject to the constraints

$$\begin{aligned} x_1 \ + 2x_2 + x_3 &= 20 \\ 2x_1 + 2x_2 + x_4 &= 30 \\ 2x_1 + \ x_2 \ + x_5 &= 25 \end{aligned}$$

and

$$x_1, x_2, x_3, x_4, x_5 \geq 0.$$

The corresponding *Tucker Tableau*, in the sense that is defined in Section 4.2, is described as

$$\begin{bmatrix} x_1 & x_2 & & \\ 1 & 2 & 20 & x_3 \\ 2 & 2 & 30 & x_4 \\ 2 & 1 & 25 & x_5 \\ 200 & 150 & 0 & \end{bmatrix}.$$

In what follows, we indicate a sequence of *pivot steps* that ultimately lead to the optimal value of the objective function $P(x_1, x_2)$. The concept of *pivot* applied to a Tucker Tableau is defined in Section 4.2; the general method behind performing such a desirable choice of pivot steps is described in Section 5.3.

Here, we illustrate these concepts prior to their definition in order to provide context and motivation for how they are used. Indeed, determining a good choice of pivot steps is ultimately the role of the Simplex Method. (See Chapter 5.)

First, solve the constraint equation

$$2x_1 + x_2 + x_5 = 25$$

for x_1 in terms of x_5 and x_2. The result is

$$x_1 = -\frac{1}{2}x_5 - \frac{1}{2}x_2 + \frac{25}{2}.$$

Then, switching the roles of x_1 and x_5 yields the following canonical slack maximization problem.

- Maximize the affine linear objective function

$$P(x_5, x_2) = -100x_5 + 50x_2 + 2500$$

- Subject to the constraints

$$\begin{aligned} -\tfrac{1}{2}x_5 + \tfrac{3}{2}x_2 + x_3 &= \tfrac{15}{2} \\ -x_5 + x_2 + x_4 &= 5 \\ \tfrac{1}{2}x_5 + \tfrac{1}{2}x_2 + x_1 &= \tfrac{25}{2} \end{aligned}$$

and

$$x_1, x_2, x_3, x_4, x_5 \geq 0.$$

The corresponding Tucker Tableau is

$$\begin{bmatrix} x_5 & x_2 & & \\ -\frac{1}{2} & \frac{3}{2} & \frac{15}{2} & x_3 \\ -1 & 1 & 5 & x_4 \\ \frac{1}{2} & \frac{1}{2} & \frac{25}{2} & x_1 \\ \text{-100} & 50 & \text{-2500} & \end{bmatrix}.$$

In interchanging the roles of x_1 and x_5, a key point is that the feasible solutions of the linear programming problems are left unchanged.

Now, repeat the above procedure with respect to the constraint equation

$$-x_5 + x_2 + x_4 = 5.$$

Solving for x_2 yields

$$x_2 = x_5 - x_4 + 5.$$

Now, interchanging the role of x_2 and x_4 yields the following canonical slack maximization problem.

- Maximize the affine linear objective function

$$P(x_5, x_4) = -50x_5 - 50x_4 + 2750$$

- Subject to the constraints

$$\begin{aligned} x_5 - \tfrac{3}{2}x_4 + x_3 &= 0 \\ -x_5 + x_4 + x_2 &= 5 \\ x_5 - \tfrac{1}{2}x_4 + x_1 &= 10 \end{aligned}$$

and

$$x_1, x_2, x_3, x_4, x_5 \geq 0.$$

The corresponding Tucker Tableau is

$$\begin{bmatrix} x_5 & x_4 & & \\ 1 & -\frac{3}{2} & 0 & x_3 \\ -1 & 1 & 5 & x_2 \\ 1 & -\frac{1}{2} & 10 & x_1 \\ -50 & -50 & -2750 & \end{bmatrix}.$$

It turns out that the optimal solution is encoded within this table. To see why, first of all, the affine linear objective function

$$P(x_5, x_4) = -50x_5 - 50x_4 + 2750$$

is decreasing with increasing values of x_5 and x_4. Thus, since the aim is to maximize P, we set x_5 and x_4 to be the smallest value. That is, we set

$$x_5 = x_4 = 0.$$

This gives

$$P(0, 0) = 2750$$

as the optimal maximum value of the objective function.

Now, when

$$x_5 = x_4 = 0$$

the given constraints imply that

$$x_3 = 0, x_2 = 5 \text{ and } x_1 = 10.$$

Hence, the optimal solution is achieved when

$$x_1 = 10, x_2 = 5, x_3 = x_4 = x_5 = 0.$$

The conclusion is that the optimal value of the objective function is

$$P(x_1 = 10, x_2 = 5) = 2750.$$

4.2 The Concept of Tucker Tableau

In solving canonical maximization/minimization linear programming problems, it is helpful to organize the data needed to specify such programming problems in *Tucker Tableau*. The purpose of this section is to describe such concepts. (See Sections 4.2.1, 4.2.2 and 4.2.3.)

4.2.1 Tucker Tableau for Canonical Slack Maximization Linear Programming Problems

The *maximization Tucker Tableau* that is associated to the *slack form*

- Maximize the affine linear objective function

$$f(x_1, \ldots, x_n) = c_1x_1 + \cdots + c_nx_n - d$$

- Subject to the constraints

$$\begin{matrix} a_{11}x_1 & + \ldots + & a_{1n}x_n & + & x_{n+1} & = & b_1 \\ \vdots & \vdots & \vdots & & \vdots & & \vdots \\ a_{m1}x_1 & + \ldots + & a_{mn}x_n & + & x_{n+m} & = & b_m \end{matrix}$$

  and

$$x_1, \ldots, x_{n+m} \geq 0$$

of a given canonical maximization linear programming problem

- Maximize the affine linear objective function

$$f(x_1, \ldots, x_n) = c_1x_1 + \cdots + c_nx_n - d$$

- Subject to the constraints

$$\begin{matrix} a_{11}x_1 & + \ldots + & a_{1n}x_n & \leq & b_1 \\ \vdots & \vdots & \vdots & & \vdots \\ a_{m1}x_1 & + \ldots + & a_{mn}x_n & \leq & b_m \end{matrix}$$

  and

$$x_1, \ldots, x_n \geq 0$$

is defined to be the collection of tabular data

$$\begin{bmatrix} x_1 & \ldots & x_n & & \\ a_{11} & \ldots & a_{1n} & b_1 & x_{n+1} \\ \vdots & \vdots & \vdots & \vdots & \vdots \\ a_{m1} & \ldots & a_{mn} & b_m & x_{n+m} \\ c_1 & \ldots & c_n & d & \end{bmatrix}. \tag{4.1}$$

We also write

$$\begin{bmatrix} x_1 & \dots & x_n & & \\ a_{11} & \dots & a_{1n} & b_1 & x_{n+1} \\ \vdots & \vdots & \vdots & \vdots & \vdots \\ a_{m1} & \dots & a_{mn} & b_m & x_{n+m} \\ c_1 & \dots & c_n & d & \end{bmatrix} = \begin{bmatrix} x_1 & \dots & x_n & & \\ a_{11} & \dots & a_{1n} & a_{1n+1} & x_{n+1} \\ \vdots & \vdots & \vdots & \vdots & \vdots \\ a_{m1} & \dots & a_{mn} & a_{mn+1} & x_{n+m} \\ a_{m+11} & \dots & a_{m+1n} & a_{m+1n+1} & \end{bmatrix}. \tag{4.2}$$

In doing so, we say that the right-hand side in (4.2) above is the associated *relabelled maximization Tucker Tableau.*

The reason for such notational considerations is mostly technical. However, such notation is useful in describing the steps of the Pivot Transform Method and the Simplex Method. It is also conceptually helpful from the vantage point of *duality theory* for linear programming programs. (See Section 4.3.)

The concept of maximization Tucker Tableau, in the sense defined by (4.1), and relabelled maximization Tucker Tableau, in the sense that is defined via the right-hand side of (4.2), provide a means to encode the data that is needed to specify a canonical form maximization linear programming problem.

In the above context, we say that $x_1, \dots, x_n$ are the *horizontal* or *independent* decision variables whereas $x_{n+1}, \dots, x_{n+m}$ are the *vertical* or *dependent* decision variables. By the *basic solution* to a given maximization Tableau (4.2), we mean the solution that is obtained with all horizontal variables equal to zero $x_1 = 0, \dots, x_n = 0$.

We say that such canonical form maximization linear programming problems and their associated maximization Tucker Tableau (4.1) and (4.2) are *basic feasible* if $b_1, \dots, b_m \geq 0$.

Example 4.2. As in Example 4.1, the maximization linear programming problem

- Maximize the linear objective function

$$P(x_1, x_2) = 200x_1 + 150x_2$$

- Subject to the constraints

$$\begin{aligned} x_1 + 2x_2 &\leq 20 \\ 2x_1 + 2x_2 &\leq 30 \\ 2x_1 + x_2 &\leq 25 \end{aligned}$$

and

$$x_1, x_2 \geq 0$$

has corresponding slack formulation

- Maximize the objection function

$$P(x_1, x_2) = 200x_1 + 150x_2$$

- Subject to the constraints

$$\begin{aligned} x_1 \ + 2x_2 + x_3 &= 20 \\ 2x_1 + 2x_2 + x_4 &= 30 \\ 2x_1 + \ x_2 \ + x_5 &= 25 \end{aligned}$$

  and

$$x_1, x_2, x_3, x_4, x_5 \geq 0$$

with corresponding Tucker Tableau

$$\begin{bmatrix} x_1 & x_2 & & \\ 1 & 2 & 20 & x_3 \\ 2 & 2 & 30 & x_4 \\ 2 & 1 & 25 & x_5 \\ 200 & 150 & 0 & \end{bmatrix}.$$

4.2.2 Tucker Tableau for Canonical Slack Minimization Linear Programming Problems

The *minimization Tucker Tableau* that is associated to the *slack form*

- Minimize the affine linear objective function

$$g(y_1, \ldots, y_m) = b_1 y_1 + \cdots + b_m y_m - d$$

- Subject to the constraints

$$\begin{aligned} a_{11} y_1 + \ldots + a_{m1} y_m &= c_1 + y_{m+1} \\ \vdots \quad\quad \vdots \quad\quad \vdots \quad\quad &\quad \vdots \quad\quad \vdots \\ a_{1n} y_1 + \ldots + a_{mn} y_m &= c_n + y_{m+n} \end{aligned}$$

  and

$$y_1, \ldots, y_{n+m} \geq 0$$

of a given *canonical minimization linear programming problem*

- Minimize the affine linear objective function

$$g(y_1, \ldots, y_m) = b_1 y_1 + \cdots + b_m y_m - d$$

- Subject to the constraints

$$\begin{array}{cccc} a_{11}y_1 + \ldots + a_{m1}y_m \geq c_1 \\ \vdots \quad \vdots \quad \vdots \quad \vdots \\ a_{1n}y_1 + \ldots + a_{mn}y_m \geq c_n \end{array}$$

and

$$y_1, \ldots, y_m \geq 0$$

is defined, similar to the case of the Tucker Tableau for canonical maximization linear programming problems, see (4.1), to be the following collection of tabular data

$$\begin{bmatrix} y_1 & a_{11} & \ldots & a_{1n} & b_1 \\ \vdots & \vdots & \vdots & \vdots & \vdots \\ y_m & a_{m1} & \ldots & a_{mn} & b_m \\ & c_1 & \ldots & c_n & d \\ & y_{m+1} & \ldots & y_{n+m} & \end{bmatrix}. \tag{4.3}$$

As in (4.2), we also write

$$\begin{bmatrix} y_1 & a_{11} & \ldots & a_{1n} & b_1 \\ \vdots & \vdots & \vdots & \vdots & \vdots \\ y_m & a_{m1} & \ldots & a_{mn} & b_m \\ & c_1 & \ldots & c_n & d \\ & y_{m+1} & \ldots & y_{n+m} & \end{bmatrix} = \begin{bmatrix} y_1 & a_{11} & \ldots & a_{1n} & a_{1n+1} \\ \vdots & \vdots & \vdots & \vdots & \vdots \\ y_m & a_{m1} & \ldots & a_{mn} & a_{mn+1} \\ & a_{m+11} & \ldots & a_{m+1n} & a_{m+1n+1} \\ & y_{m+1} & \ldots & y_{n+m} & \end{bmatrix} \tag{4.4}$$

and we refer to the right-hand side in (4.4) as a *relabelled minimization Tucker Tableau*.

In the above context, we say that $y_1, \ldots, y_m$ are the *vertical* or *independent* decision variables whereas $y_{m+1}, \ldots, y_{n+m}$ are the *horizontal* or *dependent* decision variables.

Example 4.3. Consider the minimization linear programming problem

- Minimize the objective function

$$C(y_1, y_2, y_3) = 20y_1 + 30y_2 + 25y_3$$

- Subject to the constraints

$$\begin{array}{c} y_1 + 2y_2 + 2y_3 \geq 200 \\ 2y_1 + 2y_2 + y_3 \geq 150 \end{array}$$

and

$$y_1, y_2, y_3 \geq 0.$$

The slack form of this linear programming problem is

- Minimize the objective function

$$C(y_1, y_2, y_3) = 20y_1 + 30y_2 + 25y_3$$

- Subject to the constraints

$$\begin{aligned} y_1 + 2y_2 + 2y_3 &= 200 + y_4 \\ 2y_1 + 2y_2 + y_3 &= 150 + y_5 \end{aligned}$$

and

$$y_1, y_2, y_3, y_4, y_5 \geq 0.$$

The corresponding minimization Tucker Tableau is described as

$$\begin{bmatrix} y_1 & 1 & 2 & 20 \\ y_2 & 2 & 2 & 30 \\ y_3 & 2 & 1 & 25 \\ & 200 & 150 & 0 \\ & y_4 & y_5 & \end{bmatrix}.$$

4.2.3 *Negative Transpose for Canonical Slack Minimization Linear Programming Problems*

Via the operation of *negative transpose*, each minimization Tucker Tableau

$$\begin{bmatrix} y_1 & a_{11} & \dots & a_{1n} & b_1 \\ \vdots & \vdots & \vdots & \vdots & \vdots \\ y_m & a_{m1} & \dots & a_{mn} & b_m \\ & c_1 & \dots & c_n & d \\ & y_{m+1} & \dots & y_{n+m} & \end{bmatrix} \tag{4.5}$$

determines a maximization Tucker Tableau

$$\begin{bmatrix} y_1 & \dots & y_m & & \\ -a_{11} & \dots & -a_{m1} & -c_1 & y_{m+1} \\ \vdots & \vdots & \vdots & \vdots & \vdots \\ -a_{1n} & \dots & -a_{mn} & -c_n & y_{m+n} \\ -b_1 & \dots & -b_m & -d & \end{bmatrix}. \tag{4.6}$$

The canonical form maximization linear programming problem that corresponds to the maximization Tucker Tableau (4.6), which has optimal solution that determines the optimal solution of the given minimization linear programming problem that corresponds to the minimization Tucker Tableau (4.5), is

- Maximize the affine linear objective function

$$-g(y_1, \ldots, y_m) = -b_1 y_1 - \cdots - b_m y_m + d$$

- Subject to the constraints

$$\begin{array}{ccccc} -a_{11} y_1 & - & \ldots & - & a_{m1} y_m \leq -c_1 \\ \vdots & & \vdots & & \vdots \quad\quad \vdots \\ -a_{1n} y_1 & - & \ldots & - & a_{mn} y_m \leq -c_n \end{array}$$

and

$$y_1, \ldots, y_m \geq 0.$$

Such considerations, via the operation of negative transpose, are significant as they allow for solutions to canonical form minimization linear programming problems via the Simplex Method for canonical maximization linear programming problems.

It is possible to formulate directly a Simplex Method for canonical form minimization linear programming problems. However, in light of the inherent *duality* that exists within linear programming problems, see Section 4.3, we do not pursue here such direct approaches for a Simplex Method applied to minimization Tucker Tableau. Instead, we make use of the Simplex Method applied to the negative transpose of a minimization Tucker Tableau.

Example 4.4. For the minimization linear programming problem considered in Example 4.3, the minimization Tucker Tableau

$$\begin{bmatrix} y_1 & 1 & 2 & 20 \\ y_2 & 2 & 2 & 30 \\ y_3 & 2 & 1 & 25 \\ & 200 & 150 & 0 \\ & y_4 & y_5 & \end{bmatrix}$$

has negative transpose the maximization Tucker Tableau given by

$$\begin{bmatrix} y_1 & y_2 & y_3 & & \\ -1 & -2 & -2 & -200 & y_4 \\ -2 & -2 & -1 & -150 & y_5 \\ -20 & -30 & -25 & 0 & \end{bmatrix}.$$

Having defined the concept of negative transpose for a minimization Tucker Tableau (4.5), given an optimal solution of the canonical maximization linear programming problem that is associated to the corresponding maximization Tucker Tableau (4.6), an optimal solution of the given canonical minimization linear programming problem is determined. This is illustrated in Example 4.5.

Example 4.5. The maximization linear programming problem that is associated to the Tucker Tableau

$$\begin{bmatrix} y_1 & y_2 & y_3 & & \\ -1 & -2 & -2 & -200 & y_4 \\ -2 & -2 & -1 & -150 & y_5 \\ -20 & -30 & -25 & 0 & \end{bmatrix}$$

that arises in Example 4.4, turns out to have optimal maximal solution -2750 which is achieved when $y_1 = 0$, $y_2 = 50$ and $y_3 = 50$. It follows that the minimization linear programming problem that is associated to the minimization Tucker Tableau

$$\begin{bmatrix} y_1 & 1 & 2 & 20 \\ y_2 & 2 & 2 & 30 \\ y_3 & 2 & 1 & 25 \\ & 200 & 150 & 0 \\ & y_4 & y_5 & \end{bmatrix}$$

has optimal minimal solution 2750 which is achieved when $y_1 = 0$, $y_2 = 50$ and $y_3 = 50$.

4.3 Introduction to Duality Theory for Linear Programming

In this section, we discuss a concept of *duality theory* for linear programming. The idea is to view a fixed Tucker Tableau as simultaneously determining a dual pair of maximization and minimization linear programming problems. In light of this, it is of interest to study the extent to which such dual pairs of programming problems interact in terms of their respective feasible and optimal solutions. Another source of motivation for duality theory arises from the concept of *shadow prices*. By this perspective, the dual variables provide information as to the extent to which the objective function will increase under small perturbations of the decision variables.

The main points about duality theory for linear programming are illustrated in Example 4.6.

Example 4.6. Consider the maximization linear programming problem that is associated to the canonical maximization Tableau

$$\begin{bmatrix} x_1 & x_2 & & \\ 1 & 2 & 20 & x_3 \\ 2 & 2 & 30 & x_4 \\ 2 & 1 & 25 & x_5 \\ 200 & 150 & 0 & \end{bmatrix}.$$

It is formulated as

- Maximize the objective function

$$P(x_1, x_2) = 200x_1 + 150x_2$$

- Subject to the constraints

$$\begin{aligned} x_1 + 2x_2 &\leq 20 \\ 2x_1 + 2x_2 &\leq 30 \\ 2x_1 + x_2 &\leq 25 \end{aligned}$$

and

$$x_1, x_2 \geq 0.$$

The slack form of this linear programming problem is described as

- Maximize the objective function

$$P(x_1, x_2) = 200x_1 + 150x_2$$

- Subject to the constraints

$$\begin{aligned} x_1 + 2x_2 + x_3 &= 20 \\ 2x_1 + 2x_2 + x_4 &= 30 \\ 2x_1 + x_2 + x_5 &= 25 \end{aligned}$$

and

$$x_1, x_2, x_3, x_4, x_5 \geq 0.$$

At the same time, consider the minimization linear program that is associated to the canonical minimization Tableau

$$\begin{bmatrix} y_1 & 1 & 2 & 20 \\ y_2 & 2 & 2 & 30 \\ y_3 & 2 & 1 & 25 \\ & 200 & 150 & 0 \\ & y_4 & y_5 & \end{bmatrix}.$$

That is, consider the minimization linear programming problem

- Minimize the objective function

$$C(y_1, y_2, y_3) = 20y_1 + 30y_2 + 25y_3$$

- Subject to the constraints

$$\begin{aligned} y_1 + 2y_2 + 2y_3 &\geq 200 \\ 2y_1 + 2y_2 + y_3 &\geq 150 \end{aligned}$$

and

$$y_1, y_2, y_3 \geq 0.$$

The slack form of this linear programming problem is described as

- Minimize the objective function

$$C(y_1, y_2, y_3) = 20y_1 + 30y_2 + 25y_3$$

- Subject to the constraints

$$\begin{aligned} y_1 + 2y_2 + 2y_3 - y_4 &= 200 \\ 2y_1 + 2y_2 + y_3 - y_5 &= 150 \end{aligned}$$

and

$$y_1, y_2, y_3, y_4, y_5 \geq 0.$$

In Section 5.7, we address the interesting question as to the extent to which a sequence of *pivot steps* applied to the entries of the maximization Tucker Tableau

$$\begin{bmatrix} x_1 & x_2 & & \\ 1 & 2 & 20 & x_3 \\ 2 & 2 & 30 & x_4 \\ 2 & 1 & 25 & x_5 \\ 200 & 150 & 0 & \end{bmatrix}$$

solves simultaneously the respective maximization and dual minimization linear programming problems. Such phenomena provide motivation for the concept of *Dual Tucker Tableau*, see Section 4.3.1, and the *Duality Equation*, see Section 4.3.3 (especially Theorem 4.8).

4.3.1 Dual Tucker Tableau

To emphasize the *duality* that exists in linear programming problems, it is helpful to consider the *dual Tucker Tableau* that are associated to *dual pairs* of canonical linear programming problems.

Specifically, the *dual Tucker Tableau* that is associated to a *dual pair* of canonical linear programming problems

- Maximize the affine linear objective function

$$f(x_1, \ldots, x_n) = c_1x_1 + \cdots + c_nx_n - d$$

- Subject to the constraints

$$\begin{array}{ccccccc} a_{11}x_1 & + & \ldots & + & a_{1n}x_n & \leq & b_1 \\ \vdots & & \vdots & & \vdots & & \vdots \\ a_{m1}x_1 & + & \ldots & + & a_{mn}x_n & \leq & b_m \end{array}$$

  and

$$x_1, \ldots, x_n \geq 0$$

respectively,

- Minimize the affine linear objective function

$$g(y_1, \ldots, y_m) = b_1y_1 + \cdots + b_my_m - d$$

- Subject to the constraints

$$\begin{array}{ccccccc} a_{11}y_1 & + & \ldots & + & a_{m1}y_m & \geq & c_1 \\ \vdots & & \vdots & & \vdots & & \vdots \\ a_{1n}y_1 & + & \ldots & + & a_{mn}y_m & \geq & c_n \end{array}$$

  and

$$y_1, \ldots, y_m \geq 0$$

is defined to be the following collection of tabular data:

$$\begin{bmatrix} & x_1 & \ldots & x_n & & \\ y_1 & a_{11} & \ldots & a_{1n} & b_1 & x_{n+1} \\ \vdots & \vdots & \vdots & \vdots & \vdots & \vdots \\ y_m & a_{m1} & \ldots & a_{mn} & b_m & x_{n+m} \\ & c_1 & \ldots & c_n & d & \\ & y_{m+1} & \ldots & y_{m+n} & & \end{bmatrix}. \tag{4.7}$$

We also write

$$\begin{bmatrix} & x_1 & \dots & x_n & & \\ y_1 & a_{11} & \dots & a_{1n} & b_1 & x_{n+1} \\ \vdots & \vdots & \vdots & \vdots & \vdots & \vdots \\ y_m & a_{m1} & \dots & a_{mn} & b_m & x_{n+m} \\ & c_1 & \dots & c_n & d & \\ & y_{m+1} & \dots & y_{m+n} & & \end{bmatrix} = \begin{bmatrix} & x_1 & \dots & x_n & & \\ y_1 & a_{11} & \dots & a_{1n} & a_{1n+1} & x_{n+1} \\ \vdots & \vdots & \vdots & \vdots & \vdots & \vdots \\ y_m & a_{m1} & \dots & a_{mn} & a_{mn+1} & x_{n+m} \\ & a_{m+11} & \dots & a_{m+1n} & a_{m+1n+1} & \\ & y_{m+1} & \dots & y_{m+n} & & \end{bmatrix}. \quad (4.8)$$

Similar to the case of maximization Tucker Tableau (4.1) and minimization Tucker Tableau (4.3), we say that the right-hand side in (4.8) is the *relabelled dual Tucker Tableau* that is associated to the dual Tucker Tableau (4.7).

As is the situation with relabelled maximization and minimization Tucker Tableau, see (4.2) and (4.4), the usefulness for considering relabelled dual Tucker Tableau (4.8) is apparent upon considering the steps of the Simplex Method applied to dual Tucker Tableau. (See for example in Section 6.1 where such indexing conventions for relabelled Tableau are used to describe a direct implementation of the Simplex Method in Python.)

For example, in Section 5.6.3, Theorem 4.7 below is proved as a consequence of the Simplex Method applied to dual Tucker Tableau. It turns out that the converse to item (ii) can be derived as a consequence of the *duality equations* (see Corollary 4.11) for dual pairs of linear programming problems.

Theorem 4.7. *Consider the case of a dual pair of canonical form linear programming problems that are determined by a dual Tucker Tableau* (4.7). *Respectively denote the maximization/minimization objective functions as* $f(x_1, \dots, x_n)$ *and* $g(y_1, \dots, y_m)$. *The following assertions hold true.*

(i) The dual Tableau's canonical maximization linear programming problem has an optimal solution if and only if its canonical minimization linear programming problem has an optimal solution.
(ii) If $\mathbf{x}^* = (x_1^*, \dots, x_n^*)$ *is an optimal solution of the dual Tableau's canonical maximization linear programming problem and if* $\mathbf{y}^* = (y_1^*, \dots, y_m^*)$ *is an optimal solution to the respective dual canonical minimization linear programming problem, then* $f(x_1^*, \dots, x_n^*) = g(y_1^*, \dots, y_m^*)$.

As mentioned, Theorem 4.7 can be proved as an application of the Simplex Method (see Section 5.6.3). On the other hand, it is also of interest to discuss Theorem 4.7 via the viewpoint of matrix algebra. For example, the converse to item (ii) in Theorem 4.7 follows as an application of the *duality equations* that govern optimal solutions to dual pairs of linear programming problems (see Corollary 4.11). Developing this view point via the theory of matrix algebra is the aim of Section 4.3.2 below.

4.3.2 Matrix Formulation of Dual Canonical Tableau

Here, we give a matrix formulation of dual canonical tableau. Essentially, this is only a notational reformulation but there is a conceptual benefit which arises by analogy with the duality theory which arises within the contexts of linear algebra and functional analysis.

With this aim in mind, given the data of a dual Tucker Tableau

$$\begin{bmatrix} & x_1 & \dots & x_n & & \\ y_1 & a_{11} & \dots & a_{1n} & b_1 & x_{n+1} \\ \vdots & \vdots & \vdots & \vdots & \vdots & \vdots \\ y_m & a_{m1} & \dots & a_{mn} & b_m & x_{n+m} \\ & c_1 & \dots & c_n & d & \\ & y_{m+1} & \dots & y_{m+n} & & \end{bmatrix} = \begin{bmatrix} & x_1 & \dots & x_n & & \\ y_1 & a_{11} & \dots & a_{1n} & a_{1n+1} & x_{n+1} \\ \vdots & \vdots & \vdots & \vdots & \vdots & \vdots \\ y_m & a_{m1} & \dots & a_{mn} & a_{mn+1} & x_{n+m} \\ & a_{m+11} & \dots & a_{m+1n} & a_{m+1n+1} & \\ & y_{m+1} & \dots & y_{m+n} & & \end{bmatrix} \tag{4.9}$$

define the following collection of matrices:

$$A = \begin{bmatrix} a_{11} & \dots & a_{1n} \\ \vdots & & \vdots \\ a_{m1} & \dots & a_{mn} \end{bmatrix} \in \mathbb{R}^{m \times n}$$

$$\mathbf{b} = \begin{bmatrix} b_1 \\ \vdots \\ b_m \end{bmatrix} \in \mathbb{R}^{m \times 1}, \mathbf{c} = \begin{bmatrix} c_1 & \dots & c_n \end{bmatrix} \in \mathbb{R}^{1 \times n}$$

$$\mathbf{d} = \begin{bmatrix} d \end{bmatrix} \in \mathbb{R}^{1 \times 1}, \mathbf{x} = \begin{bmatrix} x_1 \\ \vdots \\ x_n \end{bmatrix} \in \mathbb{R}^{n \times 1}, \mathbf{y} = \begin{bmatrix} y_1 & \dots & y_m \end{bmatrix} \in \mathbb{R}^{1 \times m}$$

$$\mathbf{t} = \begin{bmatrix} x_{n+1} \\ \vdots \\ x_{n+m} \end{bmatrix} \in \mathbb{R}^{m \times 1} \text{ and } \mathbf{s} = \begin{bmatrix} y_{m+1} & \cdots & y_{m+n} \end{bmatrix} \in \mathbb{R}^{1 \times n}.$$

By this notation, the matrix reformulation of the slack form of the corresponding dual canonical form linear programming problem, that is determined by the dual Tucker Tableau (4.9), is given by

- Maximize the affine linear objective function

$$\mathbf{f} = \mathbf{cx} - \mathbf{d}$$

- Subject to the constraints

$$A\mathbf{x} + \mathbf{t} = \mathbf{b}$$

  and

$$\mathbf{x}, \mathbf{t} \geq 0$$

  respectively,
- Minimize the affine linear objective function

$$\mathbf{g} = \mathbf{y}\mathbf{b} - \mathbf{d}$$

- Subject to the constraints

$$\mathbf{y}A = \mathbf{s} + \mathbf{c}$$

  and

$$\mathbf{y}, \mathbf{s} \geq 0.$$

That such matrix equations do indeed recover the given dual pair of linear programming problems follows immediately upon expanding out the given matrix multiplications.

Here, the condition that a given matrix is not negative means that all of its entries are nonnegative. Finally, the respective objective functions $f(x_1, \ldots, x_n)$ and $g(y_1, \ldots, y_m)$ are regarded as the respective 1×1 matrices **f** and **g**.

4.3.3 Statement and Proof of the Linear Programming Duality Equation

Aside from notational utility for the above formulation of dual pair linear programming problems, a key application of this viewpoint starts with *Tucker's duality equation*.

Theorem 4.8 (The Duality Equation). *For each pair of feasible solutions*

$$(\mathbf{x} = (x_1, \ldots, x_n), \mathbf{y} = (y_1, \ldots, y_m)) \in \mathbb{R}^n \times \mathbb{R}^m$$

of dual canonical linear programming problems, with corresponding dual Tucker Tableau (4.9)*, it holds true that*

$$\mathbf{g} - \mathbf{f} = \mathbf{s}\mathbf{x} + \mathbf{y}\mathbf{t}.$$

Proof. The matrix reformulation of the slack form of the given dual pair of canonical linear programming problems yields the equations

$$\begin{aligned}\mathbf{f} &= \mathbf{cx} - \mathbf{d}\\ \mathbf{b} &= A\mathbf{x} + \mathbf{t}\\ \mathbf{g} &= \mathbf{yb} - \mathbf{d}\\ \mathbf{c} &= \mathbf{y}A - \mathbf{s}.\end{aligned}$$

We then observe that

$$\begin{aligned}\mathbf{g} - \mathbf{f} &= (\mathbf{yb} - \mathbf{d}) - (\mathbf{cx} - \mathbf{d})\\ &= \mathbf{yb} - \mathbf{cx}\\ &= \mathbf{y}(A\mathbf{x} + \mathbf{t}) - (\mathbf{y}A - \mathbf{s})\mathbf{x}\\ &= \mathbf{y}A\mathbf{x} + \mathbf{yt} - \mathbf{y}A\mathbf{x} + \mathbf{sx}\\ &= \mathbf{yt} + \mathbf{sx}\\ &= \mathbf{sx} + \mathbf{yt}.\end{aligned}$$

□

This concept of linear programming duality from the viewpoint of the Simplex Method is revisited in Section 5.7. On the other hand, in Section 4.3.4 several more immediate consequences of the duality equation are derived.

4.3.4 Consequences of the Duality Equation

Here, we state and prove several consequences of the duality equation. They are of an independent interest. For example, Corollary 4.11 establishes the converse of item (ii) in Theorem 4.7. In Section 4.3.5, the question of using the duality equation to characterize optimal solutions is addressed. (See Theorem 4.12.)

Corollary 4.9. *For each pair of feasible solutions*

$$(\mathbf{x}, \mathbf{y}) \in \mathbb{R}^n \times \mathbb{R}^m$$

of a given dual canonical linear programming problem, it holds true that

$$g(y_1, \ldots, y_m) \geq f(x_1, \ldots, x_n).$$

Proof. Since the solutions of the dual canonical linear programming problems are feasible, each entry of **s**, **x**, **y** and **t** are nonnegative. Thus

$$\mathbf{sx} + \mathbf{yt} \geq 0;$$

whence, by the duality equation

$$\mathbf{g} - \mathbf{f} = \mathbf{sx} + \mathbf{yt} \geq 0.$$

□

Corollary 4.10.

(i) If a canonical maximization linear programming problem is unbounded, then the dual canonical minimization linear programming problem is not feasible.
(ii) If a canonical minimization linear programming problem is unbounded, then the dual canonical maximization linear programming problem is not feasible.

Proof. For (i), assume by contradiction that the maximization linear programming problem is unbounded and that the dual minimization linear programming problem has a feasible solution. By Corollary 4.10, this feasible solution must yield a value for the minimization objective function g that is at least the value of the maximization objective function f at each feasible solution to the given maximization linear programming problem.

However, since the maximization problem is not bounded, feasible solutions à priori exist for which $f \to \infty$. Hence no such feasible solution corresponding to a value for g can exist and the minimization linear programming problem is not feasible. The conclusion (ii) follows by replacing f and g, respectively, by $-g$ and $-f$ in (i). □

Corollary 4.11. *Each pair of feasible solutions*

$$(\mathbf{x}, \mathbf{y}) \in \mathbb{R}^n \times \mathbb{R}^m$$

of a given dual canonical linear programming problem which have the property that

$$f(x_1, \ldots, x_n) = g(y_1, \ldots, y_m)$$

are in fact optimal solutions.

Proof. No other feasible solution can increase the value of the maximization objective function f. Indeed, suppose that this was the case. Then, at each such point $f > g$ which yields a contraction to Corollary 4.9. Hence, the given feasible solutions do in fact give rise to optimal solutions. □

4.3.5 Complementary Slackness and Characterization of Optimal Solutions

As an application of the duality equation, in what follows, we seek to characterize when a pair of feasible solutions of a given dual canonical linear programming problem are optimal. To achieve this, see Theorem 4.12, we first define what is meant for a dual pair of feasible solutions to exhibit *complementary slackness*.

A pair of feasible solutions

$$(\mathbf{x} = (x_1, \ldots, x_n), \mathbf{y} = (y_1, \ldots, y_m)) \in \mathbb{R}^n \times \mathbb{R}^m$$

to a dual canonical linear programming problem, with corresponding dual Tucker Tableau (4.9), is said to exhibit *complementary slackness* if the following two conditions hold true

(i) $x_j \neq 0$ implies that $y_{m+j} = 0$ for $j = 1, \ldots, n$; and
(ii) $y_i \neq 0$ implies that $x_{n+i} = 0$ for $i = 1, \ldots, m$.

Theorem 4.12. *A pair of feasible solutions*

$$(\mathbf{x}, \mathbf{y}) \in \mathbb{R}^n \times \mathbb{R}^m$$

of dual canonical linear programming problems exhibit complementary slackness if and only if they are optimal solutions.

Proof. Assume that a pair of feasible solutions of dual canonical linear programming problems exhibit complementary slackness. Then

$$y_{m+j}x_j = 0 \text{ for } j = 1, \ldots, n$$

and

$$y_i x_{n+i} = 0 \text{ for } i = 1, \ldots, m.$$

Thus

$$\mathbf{g} - \mathbf{f} = \mathbf{sx} + \mathbf{yt} = \mathbf{0}$$

whence

$$\mathbf{g} = \mathbf{f};$$

the conclusion by Corollary 4.11 is then that the feasible solutions are optimal.

Assume given a pair of optimal solutions of dual canonical linear programming problems. Then $\mathbf{f} = \mathbf{g}$ at these solutions. Hence the duality equation implies that

$$\mathbf{sx} + \mathbf{yt} = \mathbf{0}.$$

Now since each entry of the matrices $\mathbf{s}$, $\mathbf{x}$, $\mathbf{y}$ and $\mathbf{t}$ are nonnegative it follows that

$$\mathbf{sx} = \mathbf{yt} = 0.$$

In other words,

$$\mathbf{sx} = \sum_{j=1}^{n} y_{m+j}x_j = 0 \text{ and } \mathbf{yt} = \sum_{i=1}^{m} y_i x_{n+i} = 0.$$

Again, since y_{m+j}, x_j, y_i and $x_{n+i} \geq 0$ for all i and j, it must be that at least one of the factors in each term of each summation must be zero. This is equivalent to the condition of complementary slackness. □

Exercises

4.1 Consider the minimization linear programming problem

- Minimize the linear objective function

$$g(y_1, y_2, y_3) = 10y_1 + 6y_2 + 8y_3$$

- Subject to the constraints

$$\begin{aligned} y_1 + y_2 + 2y_3 &\geq 2 \\ 5y_1 + 3y_2 + 2y_3 &\geq 1 \end{aligned}$$

and

$$y_1, y_2, y_3 \geq 0.$$

(i) Determine the corresponding slack linear programming problem.
(ii) Determine the corresponding minimization Tableau.
(iii) Determine the corresponding negative transpose maximization Tableau.

4.2 Consider the dual Tableau

$$\begin{bmatrix} & x_1 & x_2 & & \\ y_1 & a_{11} & a_{12} & b_1 & x_3 \\ y_2 & a_{21} & a_{22} & b_2 & x_4 \\ y_3 & a_{31} & a_{32} & b_3 & x_5 \\ & c_1 & c_2 & d & \\ & y_4 & y_5 & & \end{bmatrix}.$$

(i) Derive the duality equation

$$y_4x_1 + y_5x_2 + y_1x_3 + y_2x_4 + y_3x_5 = b_1y_1 + b_2y_2 + b_3y_3 - c_1x_1 - c_2x_2.$$

(ii) Determine the matrix formulation of the dual linear programming problem.

4.3 Consider the dual Tableau

$$\begin{bmatrix} & x_1 & x_2 & & \\ y_1 & 1 & 2 & 20 & x_5 \\ y_2 & 2 & 2 & 30 & x_4 \\ y_3 & 2 & 1 & 25 & x_5 \\ & 200 & 150 & 0 & \\ & y_4 & y_5 & & \end{bmatrix}.$$

(i) Verify, by explicit computation, the conclusion of the duality equation.
(ii) Is it reasonable to expect that the pair of solutions $x_1 = 10$, $x_2 = 5$, $x_3 = x_4 = x_5 = 0$ and $y_1 = 0$, $y_2 = y_3 = 50$, $y_4 = y_5 = 0$ give rise to a pair of optimal solutions for the respective dual maximization and minimization linear programming problems? Why or why not?

Chapter 5
Simplex Method via Tucker Tableau

Overview

- Tucker Tableau for the Simplex Method.
- The Pivot Transform Method for Tucker Tableau.
- The Simplex Method via the theory of Tucker Tableau.

5.1 Illustrative Example

Example 5.1 below outlines the main ideas that are involved in the Simplex Method. Precise details of each of the steps that are employed are described in Section 5.3.

Example 5.1. We indicate the key points that underlie the Simplex Method for basic feasible tableau applied to the maximization linear programming problem

- Maximize the objective function

$$f(x_1, x_2) = 200x_1 + 150x_2$$

- Subject to the constraints

$$\begin{aligned} x_1 + x_2 &\leq 20 \\ 2x_1 + 2x_2 &\leq 30 \\ 2x_1 + x_2 &\leq 25 \end{aligned}$$

and

$$x_1, x_2 \geq 0.$$

First, introduce the slack variables x_3, x_4, x_5. The corresponding canonical slack maximization linear programming problem seeks to maximize the objective function

$$f(x_1, x_2) = 200x_1 + 150x_2$$

N. Grieve, *An Introduction to Mathematical Programming and Network Science*, Springer Undergraduate Texts in Mathematics and Technology,
https://doi.org/10.1007/978-3-032-13330-4_5

subject to the constraints

$$\begin{aligned} x_1 + 2x_2 + x_3 &= 20 \\ 2x_1 + 2x_2 + x_4 &= 30 \\ 2x_1 + x_2 + x_5 &= 25 \end{aligned}$$

and

$$x_1, \ldots, x_5 \geq 0.$$

This canonical slack maximization linear programming problem has tableau

$$\begin{bmatrix} x_1 & x_2 & & \\ 1 & 2 & 20 & x_3 \\ 2 & 2 & 30 & x_4 \\ 2 & 1 & 25 & x_5 \\ 200 & 150 & 0 & \end{bmatrix}. \tag{5.1}$$

We say that the *independent*, or *horizontal*, variables are x_1 and x_2, whereas the *dependent*, or *vertical* variables are x_3, x_4,and x_5; the objective function can then be regarded as a function of x_1, x_2, x_3, x_4 and x_5.

We now discuss the key insight for pivoting the above tableau (5.1) with respect to the entry $a_{31} = 2$.

The idea is to solve for x_1 in the constraint equation

$$2x_1 + x_2 - 25 = -x_5.$$

This gives

$$x_1 = -\frac{1}{2}x_5 - \frac{1}{2}x_2 + \frac{25}{2};$$

that is

$$\frac{1}{2}x_5 + \frac{1}{2}x_2 + x_1 = \frac{25}{2}.$$

Back substitution into all of the slack constraints then gives rise to the following slack maximization linear programming problem.

- Maximize the affine linear objective function

$$f(x_5, x_2) = -100x_5 + 50x_2 + 2500$$

- Subject to the constraints

$$\begin{aligned} -\tfrac{1}{2}x_5 + \tfrac{3}{2}x_2 + x_3 &= \tfrac{15}{2} \\ -x_5 + x_2 + x_4 &= 5 \\ \tfrac{1}{2}x_5 + \tfrac{1}{2}x_2 + x_1 &= \tfrac{25}{2} \end{aligned}$$

and

$$x_5, x_2, x_3, x_4, x_1 \geq 0.$$

This slack maximization linear programming problem has tableau

$$\begin{bmatrix} x_5 & x_2 & & \\ -\frac{1}{2} & \frac{3}{2} & \frac{15}{2} & x_3 \\ -1 & 1 & 5 & x_4 \\ \frac{1}{2} & \frac{1}{2} & \frac{25}{2} & x_1 \\ -100 & 50 & -2500 & \end{bmatrix}. \tag{5.2}$$

The tableau (5.2) is obtained from the tableau (5.1) by pivoting along the tableau (5.1)'s entry $a_{31} = 2$.

We now repeat the process just considered. In doing so, we use a suitable pivot to create a new tableau from the tableau (5.2).

First consider the constraint

$$-x_5 + x_2 + x_4 = 5.$$

Solving for x_2 yields

$$x_2 = x_5 - x_4 + 5$$

which can be rewritten as

$$-x_5 + x_4 + x_2 = 5.$$

Back substituting for x_2 as described above yields the slack maximization linear programming problem.

- Maximize the affine linear objective function

$$f(x_5, x_4) = -50x_5 - 50x_4 + 2750$$

- Subject to the constraints

$$\begin{aligned} x_5 \; - \tfrac{3}{2}x_4 + x_3 &= 0 \\ -x_5 + \; x_4 \; + x_2 &= 5 \\ x_5 \; - \tfrac{1}{2}x_4 + x_1 &= 10 \end{aligned}$$

and

$$x_5, x_4, x_3, x_2, x_1 \geq 0.$$

The corresponding tableau is

$$\begin{bmatrix} x_5 & x_4 & & \\ 1 & -\frac{3}{2} & 0 & x_3 \\ -1 & 1 & 5 & x_2 \\ 1 & -\frac{1}{2} & 10 & x_1 \\ -50 & -50 & -2750 & \end{bmatrix}. \tag{5.3}$$

This tableau (5.3) was obtained from the tableau (5.2) by pivoting along the tableau (5.2)'s entry $a_{22} = 1$.

The feasible solutions of the given slack maximization linear programming problem remain unchanged under these pivoting operations. In fact, an optimal solution can be extracted from the tableau (5.3); the Simplex Method applied to the originally given slack maximization linear programming problem terminates at this stage.

Using the tableau (5.3), an optimal solution is obtained by first setting

$$x_5 = 0 \text{ and } x_4 = 0$$

and then deducing from this tableau that

$$x_3 = 0, x_1 = 10 \text{ and } x_2 = 5.$$

Finally, using the tableau (5.3), when x_1, x_2, x_3, x_4 and x_5 assume these values, the value of the objective function is

$$f(x_5 = 0, x_4 = 0) = 2750 = f(x_1 = 10, x_2 = 5).$$

5.2 Tableau for the Simplex Method

In what follows, we consider canonical maximization linear programming problems. For the case of canonical minimization linear programming problems, recall that their solutions can be found by first multiplying the objective function and all constraint inequalities, aside from the nonnegativity constraints, by minus one in order to convert the given canonical minimization problem into a canonical maximization problem. (See Example 3.9.) In terms of Tucker Tableau, this operation corresponds to taking the negative transpose. (See Section 4.2.3.)

Thus, by solving this resulting canonical maximization linear programming problem, there is no loss in generality, in what follows, by treating the case of canonical maximization linear programming problems.

Indeed, consider the case of a canonical maximization linear programming problem

- Maximize the affine linear objective function

$$f(x_1, \ldots, x_n) = c_1 x_1 + \cdots + c_n x_n - d$$

- Subject to the constraints

$$\begin{matrix} a_{11}x_1 & + \ldots + & a_{1n}x_n & \leq & b_1 \\ \vdots & \vdots & \vdots & & \vdots \\ a_{m1}x_1 & + \ldots + & a_{mn}x_n & \leq & b_m \end{matrix}$$

and

$$x_1, \ldots, x_n \geq 0.$$

Recall that it may be reformulated, in slack form, as

- Maximize the affine linear objective function

$$f(x_1, \ldots, x_n) = c_1 x_1 + \cdots + c_n x_n - d$$

- Subject to the constraints

$$\begin{matrix} a_{11}x_1 & + \ldots + & a_{1n}x_n & + & x_{n+1} & = & b_1 \\ \vdots & \vdots & \vdots & & \vdots & & \vdots \\ a_{m1}x_1 & + \ldots + & a_{mn}x_n & + & x_{n+m} & = & b_m \end{matrix}$$

and

$$x_1, \ldots, x_{n+m} \geq 0.$$

We refer to such linear programming problems as *canonical slack maximization linear programming problems*. The variables $x_{n+1}, \ldots, x_{n+m}$ are called *slack variables.*

Our approach to solving canonical maximization linear programming problems is to solve the corresponding canonical slack maximization linear programming problem. We employ an abstract construct which we call *Tucker Tableau.*

The *maximization Tucker Tableau* that corresponds to a given canonical slack maximization linear programming problem is defined to be

$$\begin{bmatrix} x_1 & \dots & x_n & & \\ a_{11} & \dots & a_{1n} & b_1 & x_{n+1} \\ \vdots & \dots & \vdots & \vdots & \vdots \\ a_{m1} & \dots & a_{mn} & b_m & x_{m+n} \\ c_1 & \dots & c_n & d & \end{bmatrix}.$$

The $x_1, \dots, x_n$ are the *independent variables* and the $x_{n+1}, \dots, x_{m+n}$ are the *dependent variables*. Such a maximization Tucker Tableau is called a *maximum basic feasible tableau* if $b_1, \dots, b_m \geq 0$.

In order to describe the steps and keep track of the relevant indices that arise in the Simplex Method, we find it notationally helpful to re-express the data that underlies the Simplex Method in the following way.

Suppose given the tableau of a canonical slack maximization linear programming problem as described above. Set

$$a_{m+1j} := c_j \text{ for } 1 \leq j \leq n$$

$$a_{m+1,n+1} := d$$

and

$$a_{in+1} := b_i \text{ for } 1 \leq i \leq m.$$

Define the *relabelled maximization tableau* of a given maximization Tucker Tableau to be

$$\begin{bmatrix} x_1 & \dots & x_n & & \\ a_{11} & \dots & a_{1n} & a_{1n+1} & x_{n+1} \\ \vdots & \dots & \vdots & \vdots & \vdots \\ a_{m1} & \dots & a_{mn} & a_{mn+1} & x_{n+m} \\ a_{m+11} & \dots & a_{m+1n} & a_{m+1n+1} & \end{bmatrix}.$$

At times it is helpful to identify a given maximization Tucker Tableau with its corresponding relabelled maximization tableau. In doing so, we write

$$\begin{bmatrix} x_1 & \dots & x_n & & \\ a_{11} & \dots & a_{1n} & b_1 & x_{n+1} \\ \vdots & \dots & \vdots & \vdots & \vdots \\ a_{m1} & \dots & a_{mn} & b_m & x_{n+m} \\ c_1 & \dots & c_n & d & \end{bmatrix} = \begin{bmatrix} x_1 & \dots & x_n & & \\ a_{11} & \dots & a_{1n} & a_{1n+1} & x_{n+1} \\ \vdots & \dots & \vdots & \vdots & \vdots \\ a_{m1} & \dots & a_{mn} & a_{mn+1} & x_{n+m} \\ a_{m+11} & \dots & a_{m+1n} & a_{m+1n+1} & \end{bmatrix}.$$

5.3 Outline of the Simplex Method via Tucker Tableau

Our approach to the *Simplex Method* via Tucker Tableau is based on three key steps.

(i) The first, is the concept of Pivot Transform for canonical maximization Tucker Tableau. (See Section 5.3.1.) Essentially, this method transforms one canonical maximization linear programming problem into another.
(ii) The second, is the Simplex Method for *maximal basic feasible Tableau*. (See Section 5.3.2.) The idea is to find an optimal solution to the given basic feasible maximization linear programming problem by performing a judicious choice of pivot transformations.
(iii) The third, is the Simplex Method for canonical maximization Tucker Tableau. (See Section 5.3.3.) The idea here is to use a suitable sequence of pivot steps to first transform, if necessary, a non-basic feasible maximal tableau into a maximal basic feasible tableau. This allows for the given slack maximization linear programming problem to be solved via the Simplex Method for maximal basic feasible Tableau.

5.3.1 Pivot Transform Method for Maximization Tucker Tableau

In this section, we describe the technical core to the Simplex Method which is achieved via the *Pivot Transform Method*. The Pivot Transform Method is applied to a given maximization Tucker Tableau which is expressed as a relabelled maximization Tucker Tableau

$$\begin{bmatrix} x_1 & \dots & x_n & & \\ a_{11} & \dots & a_{1n} & b_1 & x_{n+1} \\ \vdots & \vdots & \vdots & \vdots & \vdots \\ a_{m1} & \dots & a_{mn} & b_m & x_{n+m} \\ c_1 & \dots & c_n & d & \end{bmatrix} = \begin{bmatrix} x_1 & \dots & x_n & & \\ a_{11} & \dots & a_{1n} & a_{1n+1} & x_{n+1} \\ \vdots & \vdots & \vdots & \vdots & \vdots \\ a_{m1} & \dots & a_{mn} & a_{mn+1} & x_{n+m} \\ a_{m+11} & \dots & a_{m+1n} & a_{m+1n+1} & \end{bmatrix}. \tag{5.4}$$

The Pivot Transform Method, while technical, is otherwise elementary. The main conceptual point is that given a maximization Tucker Tableau as above, the operation of pivoting on a nonzero entry $0 \neq a_{ij}$, where $i = 1, \dots, m$ and $j = 1, \dots, n$, is equivalent to solving the ith equation of the Tableau for the jth variable and then replacing each occurrence of this variable in the other equations of the tableau by the resulting expression.

Such considerations give rise to the Pivot Transform Method. This is described below.

Pivot Transform Method

Given a maximization Tucker Tableau expressed as a relabeled tableau (5.4) together with a choice of nonzero *pivot entry*

$$p^* = p = a_{ij} \neq 0,$$

for some $1 \leq i \leq m$ and $1 \leq j \leq n$, by *pivot transform* we mean the maximization tableau

$$\begin{bmatrix} x_1 & \ldots & x_n & & \\ a'_{11} & \ldots & a'_{1n} & b'_1 & x_{n+1} \\ \vdots & \vdots & \vdots & \vdots & \vdots \\ a'_{m1} & \ldots & a'_{mn} & b'_m & x_{n+m} \\ c'_1 & \ldots & c'_n & d' & \end{bmatrix} = \begin{bmatrix} x_1 & \ldots & x_n & & \\ a'_{11} & \ldots & a'_{1n} & a'_{1n+1} & x_{n+1} \\ \vdots & \vdots & \vdots & \vdots & \vdots \\ a'_{m1} & \ldots & a'_{mn} & a'_{mn+1} & x_{n+m} \\ a'_{m+11} & \ldots & a'_{m+1n} & a'_{m+1n+1} & \end{bmatrix} \tag{5.5}$$

that is obtained as output of the *Pivot Transform Method*. This method consists of the following five steps.

(i) Replace x_j by x_{n+i} and x_{n+i} by x_j.
(ii) Replace $p^* = p = a_{ij}$ by $1/p = 1/a_{ij}$.
(iii) Replace each $a_{i\ell}$ for $\ell = 1, \ldots, \hat{j}, \ldots, n+1$ by $a_{i\ell}/a_{ij}$.
(iv) Replace each a_{kj}, $k = 1, \ldots, \hat{i}, \ldots, m+1$ by $-a_{kj}/a_{ij}$.
(v) For each $a_{k\ell}$ for $k = 1, \ldots, \hat{i}, \ldots, m+1$ and $\ell = 1, \ldots, \hat{j}, \ldots, n+1$ replace $a_{k\ell}$ by $(a_{ij}a_{k\ell} - a_{i\ell}a_{kj})/a_{ij}$.

Here, the notation $\hat{i}$, $\hat{j}$ means that the ith and jth indices are respectively omitted.

The output of the Pivot Transform Method is a maximization tableau which has corresponding canonical slack maximization linear programming problem with horizontal (that is independent) variables $x_1, \ldots, x_{n+i}, \ldots, x_n$ and vertical (that is dependent) variables $x_{n+1}, \ldots, x_j, \ldots, x_{n+m}$.

We now illustrate the Pivot Transform Method in a collection of examples. (See Examples 5.2–5.5.)

Example 5.2. Given the maximization Tucker Tableau

$$\begin{bmatrix} x_1 & x_2 & & \\ 1 & 2 & 3 & x_3 \\ 4 & 5 & 6 & x_4 \\ 7 & 8 & 9 & \end{bmatrix} = \begin{bmatrix} x_1 & x_2 & & \\ a_{11} & a_{12} & a_{13} & x_3 \\ a_{21} & a_{22} & a_{23} & x_4 \\ a_{31} & a_{32} & a_{33} & \end{bmatrix}$$

pivot on the entry

$$p^* = p = a_{22} = 5.$$

We employ each of the steps indicated in the Pivot Transform Method for pivoting with respect to $p = p^* = a_{22}$.

- $x_2 \leftrightarrow x_4$; $a_{22} \to \frac{1}{a_{22}} = \frac{1}{5}$;
- $a_{21} \to \frac{a_{21}}{5} = \frac{4}{5}$; $a_{23} \to \frac{a_{23}}{5} = \frac{6}{5}$;
- $a_{12} \to -\frac{a_{12}}{5} = -\frac{2}{5}$; $a_{32} \to -\frac{a_{32}}{5} = -\frac{8}{5}$;
- $a_{11} \to \frac{a_{22}a_{11}-a_{21}a_{12}}{a_{22}} = -\frac{3}{5}$; $a_{31} \to \frac{a_{22}a_{31}-a_{21}a_{32}}{a_{22}} = \frac{4}{5}$; $a_{13} \to \frac{a_{22}a_{13}-a_{23}a_{12}}{a_{22}} = \frac{3}{5}$; $a_{33} \to \frac{a_{22}a_{33}-a_{23}a_{32}}{a_{22}} = -\frac{3}{5}$.

The end result is

$$\begin{bmatrix} x_1 & x_4 & & \\ -\frac{3}{5} & -\frac{2}{5} & \frac{3}{5} & x_3 \\ \frac{4}{5} & \frac{1}{5} & \frac{6}{5} & x_2 \\ \frac{4}{5} & -\frac{8}{5} & -\frac{3}{5} & \end{bmatrix} = \begin{bmatrix} x_1 & x_4 & & \\ a_{11} & a_{12} & a_{13} & x_3 \\ a_{21} & a_{22} & a_{23} & x_2 \\ a_{31} & a_{32} & a_{33} & \end{bmatrix}.$$

Example 5.3 below illustrates, in a special case, the effect of performing the Pivot Transform Method on a given 2×2 canonical slack maximization linear programming problem. The general case is only notationally more complicated. (See Examples 5.4 and 5.5.)

Example 5.3. Consider a general 2×2 canonical slack maximization linear programming problem

- Maximize the affine linear objective function

$$f(x_1, x_2) = c_1x_1 + c_2x_2 - d$$

- Subject to the constraints

$$\begin{aligned} a_{11}x_1 + a_{12}x_2 + x_3 &= b_1 \\ a_{21}x_1 + a_{22}x_2 + x_4 &= b_2 \end{aligned}$$

and

$$x_1, \ldots, x_4 \geq 0.$$

Assume that $a_{11} \neq 0$. We illustrate here the effect of pivoting on

$$0 \neq p = p^* = a_{11}.$$

A consequence of the first constraint equation is that

$$x_3 = -a_{11}x_1 - a_{12}x_2 + b_1$$

and

$$x_1 = -\frac{a_{12}}{a_{11}}x_2 - \frac{1}{a_{11}}x_3 + \frac{b_1}{a_{11}}.$$

Thus, the first constraint equation can be re-expressed as

$$-\frac{a_{12}a_{11}}{a_{11}}x_2 - \frac{a_{11}}{a_{11}}x_3 + \frac{b_1 a_{11}}{a_{11}} + a_{12}x_2 - a_{11}x_1 - a_{12}x_2 + b_1 = b_1.$$

Simplifying and then dividing both sides by $-a_{11}$ (which does not change the solution set) yields the relation

$$\frac{1}{a_{11}}x_3 + \frac{a_{12}}{a_{11}}x_2 + x_1 = \frac{b_1}{a_{11}}.$$

Similarly, the second constraint equation can be re-expressed as

$$-a_{21}\frac{a_{12}}{a_{11}}x_2 - \frac{a_{21}}{a_{11}}x_3 + \frac{a_{21}b_1}{a_{11}} + a_{22}x_2 + x_4 = b_2.$$

This equation can be rewritten in the following form:

$$-\frac{a_{21}}{a_{11}}x_3 + \left(\frac{a_{11}a_{22} - a_{21}a_{12}}{a_{11}}\right)x_2 = \frac{a_{11}b_2 - a_{21}b_1}{a_{11}}.$$

Finally, the objective function

$$f(x_1, x_2) = c_1x_1 + c_2x_2 - d$$

can be expressed as

$$f(x_1, x_2) = -\frac{c_1a_{12}}{a_{11}}x_2 - \frac{c_1}{a_{11}}x_3 + \frac{c_1b_1}{a_{11}} + c_2x_2 - d.$$

Rearranging this expression yields

$$f(x_3, x_2) = -\frac{c_1}{a_{11}}x_3 + \left(\frac{a_{11}c_2 - c_1a_{12}}{a_{11}}\right)x_2 + \left(\frac{c_1b_1 - a_{11}d}{a_{11}}\right).$$

Considering Example 5.3, note that the feasible region and the objective function of the given slack form canonical maximization linear programming problem remain unchanged after the Pivot Transform Method.

Such considerations hold true in the general case. The notation is only more complicated. This is illustrated in Example 5.4 below.

Example 5.4. Here, we describe the effect of pivoting on a nonzero pivot entry

$$0 \neq p = p^* = a_{ij}$$

for some $i = 1, \ldots, m$ and $j = 1, \ldots, n$ applied to an arbitrary $m \times n$ slack form canonical maximization linear programming problem

- Maximize the affine linear objective function

$$f(x_1, \ldots, x_n) = c_1x_1 + \cdots + c_nx_n - d$$

- Subject to the constraints

$$\begin{array}{ccccccccc} a_{11}x_1 & + & \ldots & + & a_{1n}x_n & + & x_{n+1} & = & b_1 \\ \vdots & & \vdots & & \vdots & & \vdots & & \vdots \\ a_{i1}x_1 & + & \ldots & + & a_{in}x_n & + & x_{n+i} & = & b_i \\ \vdots & & \vdots & & \vdots & & \vdots & & \vdots \\ a_{m1}x_1 & + & \ldots & + & a_{mn}x_n & + & x_{n+m} & = & b_m \end{array}$$

  and

$$x_1, \ldots, x_{n+m} \geq 0.$$

To begin with, in row i, write x_j in terms of $x_1, \ldots, \hat{x}_j, \ldots, x_n$ and x_{n+i}. This gives

$$x_j = -\frac{a_{i1}}{a_{ij}}x_1 - \cdots - \frac{a_{ij-1}}{a_{ij}}x_{j-1} - \frac{a_{ij+1}}{a_{ij}}x_{j+1} - \cdots - \frac{a_{in}}{a_{ij}}x_n - \frac{1}{a_{ij}}x_{n+i} + \frac{b_i}{a_{ij}}.$$

Similarly, write x_{n+i} in terms of $x_1, \ldots, x_n$. This gives

$$x_{n+i} = -a_{i1}x_1 - \cdots - a_{in}x_n + b_i.$$

Using these above relations, the conclusion is that the ith constraint can be re-expressed, upon simplifying and dividing by $-a_{ij}$, as

$$\frac{a_{11}}{a_{ij}}x_1 + \cdots + \frac{1}{a_{ij}}x_{n+i} + \cdots + \frac{a_{1n}}{a_{ij}}x_n + x_j = \frac{b_i}{a_{ij}}.$$

Likewise, the kth constraint, with $k \neq i$, can be re-expressed as

$$\frac{(a_{ij}a_{k1} - a_{i1}a_{kj})}{a_{ij}}x_1 + \cdots - \frac{a_{kj}}{a_{ij}}x_{n+i} + \cdots + \frac{(a_{ij}a_{kn} - a_{in}a_{kj})}{a_{ij}}x_n = \frac{a_{ij}b_k - a_{kj}b_i}{a_{ij}}.$$

Finally, the objective function

$$f(x_1, \ldots, x_n) = c_1x_1 + \cdots + c_nx_n - d$$

can be re-expressed as

$$f(x_1, \ldots, x_{j-1}, x_{n+i}, x_{j+1}, \ldots, x_n) = \\ \frac{(a_{ij}c_1 - c_j a_{i1})}{a_{ij}} x_1 + \cdots - \frac{c_j}{a_{11}} x_{n+i} + \cdots + \frac{(a_{ij}c_n - c_j a_{in})}{a_{ij}} x_n.$$

As a final illustration of the operation of the Pivot Transform Method applied to maximization Tucker Tableau, we discuss in detail, see Example 5.5, the case of pivots applied to maximization Tucker Tableau for the situation that $m = n = 2$.

Example 5.5. Consider the following canonical maximization Tucker Tableau:

$$\begin{bmatrix} x_1 & x_2 & & \\ a_{11} & a_{12} & b_1 & x_3 \\ a_{21} & a_{22} & b_2 & x_4 \\ c_1 & c_2 & d & \end{bmatrix} = \begin{bmatrix} x_1 & x_2 & & \\ a_{11} & a_{12} & a_{13} & x_3 \\ a_{21} & a_{22} & a_{23} & x_4 \\ a_{31} & a_{32} & a_{33} & \end{bmatrix}.$$

If

$$0 \neq p = p^* = a_{ij}$$

is a nonzero pivot entry, for some $i, j = 1, 2$, then, as emphasized, pivoting on a_{ij} amounts to solving the ith equation of the tableau for the jth variable and replacing each occurrence of this variable in the other equations of the tableau by the resulting expression.

Here, we describe each of the possible four cases explicitly.

- **Pivot on** a_{11}. In this case $x_1 \leftrightarrow x_3$. Further
  - $a_{11} \to 1/a_{11}$;
  - $a_{12} \to a_{12}/a_{11}$;
  - $a_{13} \to a_{13}/a_{11}$;
  - $a_{21} \to -a_{21}/a_{11}$;
  - $a_{22} \to (a_{11}a_{22} - a_{12}a_{21})/a_{11}$;
  - $a_{23} \to (a_{11}a_{23} - a_{13}a_{21})/a_{11}$;
  - $a_{31} \to -a_{31}/a_{11}$;
  - $a_{32} \to (a_{11}a_{32} - a_{12}a_{31})/a_{11}$; and
  - $a_{33} \to (a_{11}a_{33} - a_{13}a_{31})/a_{11}$.
- **Pivot on** a_{12}. In this case $x_1 \leftrightarrow x_4$. Further
  - $a_{11} \to -a_{11}/a_{12}$;
  - $a_{12} \to 1/a_{12}$;
  - $a_{13} \to -a_{13}/a_{12}$;
  - $a_{21} \to (a_{12}a_{21} - a_{11}a_{22})/a_{12}$;
  - $a_{22} \to a_{22}/a_{12}$;
  - $a_{23} \to (a_{12}a_{23} - a_{13}a_{22})/a_{12}$;

- $a_{31} \to (a_{12}a_{31} - a_{11}a_{32})/a_{12}$;
- $a_{32} \to a_{32}/a_{12}$; and
- $a_{33} \to (a_{12}a_{33} - a_{13}a_{32})/a_{12}$.

• **Pivot on** a_{21}. In this case $x_2 \leftrightarrow x_3$. Further

- $a_{11} \to -a_{11}/a_{21}$;
- $a_{12} \to (a_{21}a_{12} - a_{22}a_{11})/a_{21}$;
- $a_{13} \to (a_{21}a_{13} - a_{23}a_{11})/a_{21}$;
- $a_{21} \to 1/a_{21}$;
- $a_{22} \to a_{22}/a_{21}$;
- $a_{23} \to a_{23}/a_{21}$;
- $a_{31} \to -a_{31}/a_{21}$;
- $a_{32} \to (a_{21}a_{32} - a_{23}a_{31})/a_{21}$; and
- $a_{33} \to (a_{21}a_{33} - a_{23}a_{31})/a_{21}$.

• **Pivot on** a_{22}. In this case $x_2 \leftrightarrow x_4$. Further

- $a_{11} \to (a_{22}a_{11} - a_{12}a_{21})/a_{22}$;
- $a_{12} \to -a_{12}/a_{22}$;
- $a_{13} \to (a_{22}a_{13} - a_{23}a_{12})/a_{22}$;
- $a_{21} \to a_{21}/a_{22}$;
- $a_{22} \to 1/a_{22}$;
- $a_{23} \to a_{23}/a_{22}$;
- $a_{31} \to (a_{22}a_{31} - a_{21}a_{32})/a_{22}$;
- $a_{32} \to -a_{32}/a_{22}$; and
- $a_{33} \to (a_{22}a_{33} - a_{23}a_{32})/a_{22}$.

5.3.2 *Simplex Method for Maximum Basic Feasible Tableau*

In what follows, we first describe the *Simplex Method* applied to what we call *basic feasible maximization tableau*. Having done so, the Simplex Method applied to the case of more general Tucker Tableau reduces to the case of the Simplex Method for basic feasible tableau by a suitable choice of pivot transformations.

The formulation of the Simplex Method that we describe here includes an implementation of the so called *anti-cycling rules*, in the sense of [2]. Our implementation of the Simplex Method thus terminates.

For precise statements, we say that a maximization Tucker Tableau

$$\begin{bmatrix} x_1 & \dots & x_n & & \\ a_{11} & \dots & a_{1n} & b_1 & x_{n+1} \\ \vdots & & \vdots & \vdots & \vdots \\ a_{m1} & \dots & a_{mn} & b_m & x_{n+m} \\ c_1 & \dots & c_n & d & \end{bmatrix}$$

is a *basic feasible tableau* if it has the property that $b_1, \ldots, b_m \geq 0$.

Having defined the concept of basic feasible tableau, the Simplex Method, with anti-cycling rules, for basic feasible maximization Tucker Tableau is formulated in the following way.

Simplex Method for basic feasible maximization Tucker Tableau

Given a basic feasible maximization Tucker Tableau

$$\begin{bmatrix} x_1 & \ldots & x_n & & \\ a_{11} & \ldots & a_{1n} & b_1 & x_{n+1} \\ \vdots & \vdots & \vdots & \vdots & \vdots \\ a_{m1} & \ldots & a_{mn} & b_m & x_{n+m} \\ c_1 & \ldots & c_n & d & \end{bmatrix} = \begin{bmatrix} x_1 & \ldots & x_n & & \\ a_{11} & \ldots & a_{1n} & a_{1n+1} & x_{n+1} \\ \vdots & \vdots & \vdots & \vdots & \vdots \\ a_{m1} & \ldots & a_{mn} & a_{mn+1} & x_{n+m} \\ a_{m+11} & \ldots & a_{m+1n} & a_{m+1n+1} & \end{bmatrix}$$

so, in particular, $b_1, \ldots, b_m \geq 0$, do the following sequence of steps.

(i) If $c_1, \ldots, c_n \leq 0$, then STOP; the basic solution of the current tableau is optimal. Otherwise continue.

(ii) Set

$$j := \min_{1 \leq \ell \leq n} \{\ell : c_\ell > 0\}.$$

(iii) If $a_{1j}, \ldots, a_{mj} \leq 0$, then STOP; the maximization linear programming problem that corresponds to the tableau is unbounded.

(iv) Set

$$M := \min_{1 \leq i \leq m} \{b_i / a_{ij} : a_{ij} > 0\}$$

and define

$$i := \min_{1 \leq k \leq m} \{k : b_k / a_{kj} = M \text{ and } a_{kj} > 0\}.$$

(v) Pivot on

$$0 \neq p = p^* = a_{ij}$$

for a_{ij} uniquely determined by (ii) and (iv), and return to Step (i) for the resulting basic feasible tableau.

The output of the Simplex Method for basic feasible maximization Tucker Tableau is a maximization Tucker Tableau which has the property that its basic feasible solution gives rise to the optimal value (should one exist) of the underlying maximization linear programming problem.

Example 5.6. Here, we use the Simplex Method to determine an optimal solution to the maximization linear programming problem that is associated to the maximization Tucker Tableau

$$\begin{bmatrix} x_1 & x_2 & & \\ 2 & 1 & 8 & x_3 \\ 1 & 2 & 10 & x_4 \\ 30 & 50 & 0 & \end{bmatrix} = \begin{bmatrix} x_1 & x_2 & & \\ a_{11} & a_{12} & a_{13} & x_3 \\ a_{21} & a_{22} & a_{23} & x_4 \\ a_{31} & a_{32} & a_{33} & \end{bmatrix}.$$

This is a basic feasible tableau and hence we may thus apply the Simplex Method for basic feasible tableaux. We first need to select a pivot entry. At least one of $c_1, c_2 > 0$ and so we can take $j = 1$; $c_1 = 30$. Now, at least one of $a_{11}, a_{21} > 0$ and

$$\min_{1 \le i \le 2} \left\{ \frac{b_i}{a_{i1}} : a_{i1} > 0 \right\} = \min \left\{ \frac{8}{2}, \frac{10}{1} \right\} = 4.$$

Observe that $\frac{b_1}{a_{11}} = 4$ achieves the minimum.

We now pivot on

$$p^* = a_{11} = 2.$$

The steps involved are described in the following way.

- $x_1 \leftrightarrow x_3$;
- $a_{11} \to \frac{1}{a_{11}} = \frac{1}{2}$;
- $a_{12} \to \frac{a_{12}}{a_{11}} = \frac{1}{2}$; $a_{13} \to \frac{a_{13}}{a_{11}} = \frac{4}{2} = 2$;
- $a_{21} \to -\frac{a_{21}}{a_{11}} = -\frac{1}{2}$; $a_{31} \to -\frac{a_{31}}{a_{11}} = -15$; and
- $a_{22} \to \frac{a_{11}a_{22} - a_{12}a_{21}}{a_{11}} = \frac{3}{2}$; $a_{32} \to \frac{a_{11}a_{32} - a_{13}a_{31}}{a_{11}} = 35$; $a_{23} \to \frac{a_{11}a_{23} - a_{13}a_{21}}{a_{11}} = 6$; $a_{33} \to \frac{a_{11}a_{33} - a_{13}a_{31}}{a_{11}} = -120$.

The end result is

$$\begin{bmatrix} x_3 & x_2 & & \\ \frac{1}{2} & \frac{1}{2} & 4 & x_1 \\ -\frac{1}{2} & \frac{3}{2} & 6 & x_4 \\ -15 & 35 & -120 & \end{bmatrix} = \begin{bmatrix} x_3 & x_2 & & \\ a_{11} & a_{12} & a_{13} & x_1 \\ a_{21} & a_{22} & a_{23} & x_4 \\ a_{31} & a_{32} & a_{33} & \end{bmatrix}.$$

Now, since the objective function coefficients are all not negative we need to perform an additional pivot with respect to a suitable pivot entry. In particular, observe that

$$c_2 = 35 > 0;$$

so take $j = 2$. Note also that at least one $a_{12}, a_{22} > 0$. Thus

$$\min_{1 \le i \le 2} \left\{ \frac{b_i}{a_{i2}} : a_{i2} > 0 \right\} = \min \left\{ \frac{4}{\frac{1}{2}}, \frac{6}{\frac{3}{2}} \right\} = \min\{8, 4\}; \text{ and}$$

$$\frac{b_2}{a_{22}} = 4$$

achieves this minimum. Thus, we must now pivot on

$$p^* = a_{22} = \frac{3}{2}.$$

This is described as follows.

- $x_2 \leftrightarrow x_4$;
- $a_{22} \to \frac{2}{3}$; $a_{21} \to \frac{-\frac{1}{2}}{\frac{3}{2}} = -\frac{1}{3}$; $a_{23} \to \frac{6}{\frac{3}{2}} = 4$;
- $a_{12} \to \frac{-\frac{1}{2}}{\frac{3}{2}} = -\frac{1}{3}$; $a_{32} \to -\frac{35}{\frac{3}{2}} = -\frac{70}{3}$; and
- $a_{11} \to \frac{a_{22}a_{11}-a_{21}a_{12}}{a_{22}} = \frac{2}{3}; a_{13} \to \frac{a_{22}a_{13}-a_{23}a_{12}}{a_{22}} = 2; a_{31} \to \frac{a_{22}a_{31}-a_{21}a_{32}}{a_{22}} = -\frac{10}{3}\, a_{33} \to \frac{a_{22}a_{33}-a_{23}a_{32}}{a_{22}} = -260.$

The end result is that

$$\begin{bmatrix} x_3 & x_4 & & \\ \frac{2}{3} & -\frac{1}{3} & 2 & x_1 \\ -\frac{1}{3} & \frac{2}{3} & 4 & x_2 \\ -\frac{10}{3} & -\frac{70}{3} & -260 & \end{bmatrix} = \begin{bmatrix} x_3 & x_4 & & \\ a_{11} & a_{12} & a_{13} & x_1 \\ a_{21} & a_{22} & a_{23} & x_2 \\ a_{31} & a_{32} & a_{33} & \end{bmatrix}.$$

It follows that the optimal solution occurs when

$$x_1 = 2 \text{ and } x_2 = 4$$

in which case the objective function has value

$$P(x_1, x_2) = 260.$$

5.3.3 *Simplex Method for Maximization Linear Programming Problems*

Our formulation of the *Simplex Method* for maximization linear programming problem terminates and prevents the possibility of *cycling* in the sense that is explained in [2]. It is formulated in the following way via maximization Tucker Tableau.

Simplex Method for maximization Tucker Tableau

Given an arbitrary maximization Tucker Tableau

$$\begin{bmatrix} x_1 & \dots & x_n & & \\ a_{11} & \dots & a_{1n} & b_1 & x_{n+1} \\ \vdots & \vdots & \vdots & \vdots & \vdots \\ a_{m1} & \dots & a_{mn} & b_m & x_{n+m} \\ c_1 & \dots & c_n & d & \end{bmatrix} = \begin{bmatrix} x_1 & \dots & x_n & & \\ a_{11} & \dots & a_{1n} & a_{1n+1} & x_{n+1} \\ \vdots & \vdots & \vdots & \vdots & \vdots \\ a_{m1} & \dots & a_{mn} & a_{mn+1} & x_{n+m} \\ a_{m+11} & \dots & a_{m+1n} & a_{m+1n+1} & \end{bmatrix}$$

do the following sequence of six steps.

(i) If $b_1, \ldots, b_m \geq 0$, then the given maximization Tableau is basic feasible, go to Step (vi).

(ii) Set $i := \max_{1 \leq k \leq m}\{k : b_k < 0\}$.

(iii) If $a_{i1}, \ldots, a_{in} \geq 0$, then Stop. The given maximization linear programming problem that corresponds to the given Tableau is infeasible. Otherwise continue.

(iv) If $i = m$, then set

$$j := \min_{1 \leq \ell \leq n} \{\ell : a_{m\ell} < 0\}.$$

Pivot on

$$0 \neq p = p^* = a_{mj}$$

and go to Step (i). Otherwise continue.

(v) If $i < m$, then set

$$j := \min_{1 \leq \ell \leq n} \{\ell : a_{i\ell} < 0\}$$

and

$$M := \min_{m \geq k > i} \left\{ \{b_i/a_{ij}\} \bigcup \{b_k/a_{kj} : a_{kj} > 0\} \right\}.$$

If $b_i/a_{ij} = M$ then set $\ell := i$. Otherwise set

$$\ell := \min_{m \geq k > i} \{k : a_{kj} > 0 \text{ and } b_k/a_{kj} = M\}.$$

Pivot on

$$0 \neq p = p^* = a_{\ell j}$$

and then go to Step (i).

(vi) Apply the Simplex Method for the given basic feasible maximization Tableau.

The output of the Simplex Method for maximization Tucker Tableau is a maximization Tucker Tableau which has the property that its basic feasible solution gives rise to the optimal value (should one exist) of the underlying maximization linear programming problem.

5.3.4 Simplex Method for Minimization Linear Programming Problems

A Simplex Method for minimization linear programming problems can be formulated via the operation of negative transpose applied to minimization Tucker Tableau. This is achieved in the following way.

Simplex Method for minimization Tucker Tableau

Given an arbitrary minimization Tucker Tableau

$$\begin{bmatrix} y_1 & a_{11} & \dots & a_{1n} & b_1 \\ \vdots & \vdots & \vdots & \vdots & \vdots \\ y_m & a_{m1} & \dots & a_{mn} & b_m \\ & c_1 & \dots & c_n & d \\ & y_{m+1} & \dots & y_{n+m} & \end{bmatrix} = \begin{bmatrix} y_1 & a_{11} & \dots & a_{1n} & a_{1n+1} \\ \vdots & \vdots & \vdots & \vdots & \vdots \\ y_m & a_{m1} & \dots & a_{mn} & a_{mn+1} \\ & a_{m+11} & \dots & a_{m+1n} & a_{m+1n+1} \\ & y_{m+1} & \dots & y_{n+m} & \end{bmatrix}$$

apply the Simplex Method to the corresponding negative transpose maximization Tucker Tableau

$$\begin{bmatrix} y_1 & \dots & y_m & & \\ -a_{11} & \dots & -a_{m1} & -c_1 & y_{m+1} \\ \vdots & \vdots & \vdots & \vdots & \vdots \\ -a_{1n} & \dots & -a_{mn} & -c_n & y_{m+n} \\ -b_1 & \dots & -b_m & -d & \end{bmatrix}.$$

The output of the Simplex Method for minimization Tucker Tableau is a maximization Tucker Tableau which has the property that its basic feasible solution gives rise to the optimal value (should one exist) of the underlying minimization linear programming problem.

5.4 Simplex Method: Explicitly Worked Out Examples

Example 5.7. Consider the Tucker Tableau

$$\begin{bmatrix} x_1 & x_2 & & \\ -1 & -2 & -3 & x_3 \\ 1 & 1 & 3 & x_4 \\ 1 & 1 & 2 & x_5 \\ -2 & 4 & 0 & \end{bmatrix}$$

which corresponds to the canonical maximization linear programming problem

- Maximize the linear objective function

$$f(x_1, x_2) = -2x_1 + 4x_2$$

- Subject to the constraints

$$\begin{aligned} -x_1 - 2x_2 &\leq -3 \\ x_1 + x_2 &\leq 3 \\ x_1 + x_2 &\leq 2 \end{aligned}$$

and

$$x_1, x_2 \geq 0.$$

(Note that the constraint $x_1 + x_2 \leq 3$ is redundant.)

We want to apply the Simplex Method to the given tableau. First of all, the given tableau is not basic feasible since $b_1 = -3 < 0$.

However, applying the general Simplex Method, for slack maximization linear programming problems yields that upon pivoting on the (3, 1) entry and then on the $(1, 2)$ entry yields a basic feasible tableau.

Then applying the Simplex Method for this basic feasible tableau amounts to pivoting, respectively, on the entries (3, 1) and $(3, 2)$. The result is a tableau that contains the optimal solution $f(0, 2) = 8$.

The relevant tableau and pivot steps are described in the following way. Starting with the maximization tableau

$$\begin{bmatrix} x_1 & x_2 & & \\ -1 & -2 & -3 & x_3 \\ 1 & 1 & 3 & x_4 \\ 1 & 1 & 2 & x_5 \\ -2 & 4 & 0 & \end{bmatrix}$$

pivot on the entry

$$p^* = a_{31} = 1$$

to obtain the tableau

$$\begin{bmatrix} x_5 & x_2 & & \\ 1 & -1 & -1 & x_3 \\ -1 & 0 & 1 & x_4 \\ 1 & 1 & 2 & x_1 \\ 2 & 6 & 4 & \end{bmatrix}.$$

Pivoting on

$$p^* = a_{12} = -1$$

yields the following basic feasible tableau

$$\begin{bmatrix} x_5 & x_4 & & \\ -1 & 1 & 1 & x_2 \\ -1 & 0 & 1 & x_3 \\ 2 & 1 & 1 & x_1 \\ 8 & 6 & -2 & \end{bmatrix};$$

pivoting on

$$p^* = a_{31} = 2$$

yields the tableau

$$\begin{bmatrix} x_1 & x_3 & & \\ 1/2 & -1/2 & 3/2 & x_2 \\ 1/2 & 1/2 & 3/2 & x_4 \\ 1/2 & 1/2 & 1/2 & x_5 \\ -4 & 2 & -6 & \end{bmatrix}.$$

Finally, pivoting on

$$p^* = a_{32} = 1/2$$

yields the tableau

$$\begin{bmatrix} x_1 & x_4 & & \\ 1 & 1 & 2 & x_2 \\ 0 & -1 & 1 & x_3 \\ 1 & 2 & 1 & x_5 \\ -6 & -4 & -8 & \end{bmatrix}.$$

Using this final tableau it follows that an optimal solution is obtained as

$$f(x_1, x_2) = 8 \text{ with } x_1 = 0 \text{ and } x_2 = 2.$$

Example 5.8. Consider the following minimization Tucker Tableau:

$$\begin{bmatrix} y_1 & 20 & 25 & 300 \\ y_2 & 40 & 20 & 500 \\ & 1000 & 800 & 0 \\ & y_3 & y_4 & \end{bmatrix}.$$

We use the technique of negative transpose to find an optimal solution of the corresponding canonical minimization linear programming problem

- Minimize the linear objective function

$$g(y_1, y_2) = 300y_1 + 500y_2$$

- Subject to the constraints

$$\begin{aligned} 20y_1 + 40y_2 &\geq 1000 \\ 25y_1 + 20y_2 &\geq 800 \end{aligned}$$

and

$$y_1, y_2 \geq 0.$$

The negative transpose of this minimization tableau is

$$\begin{bmatrix} y_1 & y_2 & & \\ -20 & -40 & -1000 & y_3 \\ -25 & -20 & -800 & y_4 \\ -300 & -500 & 0 & \end{bmatrix}.$$

It corresponds to the maximization linear programming problem

- Maximize the linear objective function

$$-g(y_1, y_2) = -300y_1 - 500y_2$$

- Subject to the constraints

$$\begin{aligned} -20y_1 - 40y_2 &\leq -1000 \\ -25y_1 - 20y_2 &\leq -800 \end{aligned}$$

and

$$y_1, y_2 \geq 0.$$

This tableau is not a basic feasible tableau. So, first we need to apply the general Simplex Method (that is first perform a suitable choice of pivots to obtain a basic feasible tableau).

Pivot on

$$p^* = a_{21} = -25.$$

This gives the tableau

$$\begin{bmatrix} y_4 & y_2 & & \\ -4/5 & -24 & -360 & y_3 \\ -1/25 & 4/5 & 32 & y_1 \\ -12 & -260 & 9600 & \end{bmatrix}.$$

Applying the general form of the Simplex Method amounts to pivoting on

$$p^* = a_{12} = -24.$$

The result is the tableau

$$\begin{bmatrix} y_4 & y_1 & & \\ 1/30 & -1/24 & 15 & y_2 \\ -2/30 & 1/30 & 20 & y_1 \\ -10/3 & -260/24 & 13500 & \end{bmatrix}.$$

This is a basic feasible tableau with optimal solution

$$-g(y_1, y_2) = -13500 \text{ with } y_1 = 20 \text{ and } y_2 = 15.$$

The conclusion is that the given minimization programming problem has optimal solution

$$g(y_1, y_2) = 13500 \text{ with } y_1 = 20 \text{ and } y_2 = 15.$$

In Example 5.9 below, we revisit Example 3.3 and discuss concepts from *sensitivity analysis* within the context of the Simplex Method.

Example 5.9. Consider the maximization linear programming problem

- Maximize the objective function

$$f(x_1, x_2) = 3x_1 + 4x_2$$

- Subject to the constraints

$$\begin{aligned} x_1 + x_2 &\leq 4 \\ 3x_1 + 5x_2 &\leq 15 \end{aligned}$$

  and

$$x_1, x_2 \geq 0.$$

This programming problem has associated Tucker Tableau

$$\begin{bmatrix} x_1 & x_2 & & \\ 1 & 1 & 4 & x_3 \\ 3 & 5 & 15 & x_4 \\ 3 & 4 & 0 & \end{bmatrix}.$$

Pivoting first on the entry

$$p^* = a_{11}$$

and then on the entry

$$p^* = a_{22}$$

renders the following optimal basic feasible tableau:

$$\begin{bmatrix} x_3 & x_4 & & \\ 2.5 & -0.5 & 2.5 & x_1 \\ -1.5 & 0.5 & 1.5 & x_2 \\ -1.5 & -0.5 & -13.5 & \end{bmatrix}.$$

The conclusion given the above optimal basic feasible tableau is that the maximum value of the objective function is

$$f = 13.5$$

and that this is achieved when $x_1 = 2.5$ and $x_2 = 1.5$.

On the other hand, consider the maximization linear programming problem

- Maximize the objective function

$$f(x_1, x_2) = 3x_1 + 4x_2$$

- Subject to the constraints

$$5x_1 + 3x_2 \leq 15$$
$$2x_1 + 6x_2 \leq 12$$

and

$$x_1, x_2 \geq 0.$$

This programming problem has feasible region which can be viewed as a small perturbation of the originally given linear programming problem. It has associated Tucker Tableau

$$\begin{bmatrix} x_1 & x_2 & & \\ 5 & 3 & 15 & x_3 \\ 2 & 6 & 12 & x_4 \\ 3 & 4 & 0 & \end{bmatrix}.$$

Upon performing the above given sequence of pivot steps, that is successively pivoting on the tableau entries

$$p^* = a_{11}$$

and

$$p^* = a_{22},$$

the end result is the following optimal basic feasible maximization Tucker Tableau:

$$\begin{bmatrix} x_3 & x_4 & & \\ .025 & -0.125 & 2.25 & x_1 \\ -0.0833 & 0.20833 & 1.25 & x_2 \\ -0.41666 & -0.45833 & -11.75 & \end{bmatrix}.$$

The conclusion is that the optimal solution of the perturbed programming problem is

$$f(x_1, x_2) = 11.75.$$

It is achieved when $x_1 = 2.25$ and $x_2 = 1.25$.

One interesting feature of this example is that actually the same sequence of pivot steps simultaneously produced an optimal basic feasible maximization Tucker Tableau for each of the respective programming problems.

Such considerations illustrate the general principle of studying the effect of the Simplex Method on small perturbations of a given linear programming problem as a means to understand the sensitivity of optimal solutions in terms of small perturbations of the input parameters.

5.5 Proof of Correctness for the Simplex Method

Our approach to proving correctness of the Simplex Method follows closely the methodology of Bland [2], Balinski and Tucker [1] and Dantzig et al. [7]. A proof of correctness for the Simplex Method, as we give here can be extracted from those works. Another viewpoint is given by Chvátal in [6], whereas a more informal expository account of the Simplex Method, and proof its correctness, is given in [10]. Matrix approaches to the Simplex Method and proofs of its correctness by that perspective are given in [12] and [8] (among others).

The Simplex Method solves the given canonical maximization linear programming problem by finding an optimal solution that is an extreme point of the programming problem's feasible region. To understand how this is achieved, there are a few key points.

First of all, the effect of the Pivot Transform Method is that its output always produces a tableau which has the property that its *basic solution*—that is the solution that is obtained by setting all horizontal variables equal to zero—always corresponds to an extreme point of the feasible region. Moreover, the value of the tableau's objective function at this basic feasible solution is equal to the value of the given canonical maximization linear programming problem's objective function at the corresponding extreme point of the feasible region.

Finally, since an optimal solution is achieved at an extreme point (see Lemma 3.25 or item (i) in Theorem 3.24), the proof of correctness of the Simplex Method follows by showing that at each iteration of the Simplex Method for basic feasible maximization Tucker Tableau, the objective function does not increase (in fact it strictly decreases, see [2]) and in fact that only a finite number of iterations are required to find an extreme point that in fact maximizes the objective function.

The key technical point to the proof of correctness of the Simplex Method is given in Lemma 5.10 below. The purpose of the lemma is to relate the basic solution of a given Tucker Tableau and pivots thereof to the given programming problem's feasible region. With this lemma in hand, the proof of the Simplex Method is readily deduced.

Lemma 5.10. *Let*

$$\begin{bmatrix} x_1 & \dots & x_n & & \\ a_{11} & \dots & a_{1n} & b_1 & x_{n+1} \\ \vdots & \vdots & \vdots & \vdots & \vdots \\ a_{m1} & \dots & a_{mn} & b_m & x_{n+m} \\ c_1 & \dots & c_n & d & \end{bmatrix} = \begin{bmatrix} x_1 & \dots & x_n & & \\ a_{11} & \dots & a_{1n} & a_{1n+1} & x_{n+1} \\ \vdots & \vdots & \vdots & \vdots & \vdots \\ a_{m1} & \dots & a_{mn} & a_{mn+1} & x_{n+m} \\ a_{m+11} & \dots & a_{m+1n} & a_{m+1n+1} & \end{bmatrix} \tag{5.6}$$

be the maximization Tucker Tableau that is determined by a given canonical form maximization linear programming problem. Let S denote the programming problem's feasible region.

Let

$$\begin{bmatrix} x_1 & \dots & x_n & & \\ a'_{11} & \dots & a'_{1n} & b'_1 & x_{n+1} \\ \vdots & \vdots & \vdots & \vdots & \vdots \\ a'_{m1} & \dots & a'_{mn} & b'_m & x_{n+m} \\ c'_1 & \dots & c'_n & d' & \end{bmatrix} = \begin{bmatrix} x_1 & \dots & x_n & & \\ a'_{11} & \dots & a'_{1n} & a'_{1n+1} & x_{n+1} \\ \vdots & \vdots & \vdots & \vdots & \vdots \\ a'_{m1} & \dots & a'_{mn} & a'_{mn+1} & x_{n+m} \\ a'_{m+11} & \dots & a'_{m+1n} & a'_{m+1n+1} & \end{bmatrix} \tag{5.7}$$

be the maximization Tucker Tableau that is obtained upon pivoting on a given choice of nonzero pivot entry

$$p^* = p = a_{ij} \neq 0 \, for \, some \, 1 \leq i \leq m \, and \, 1 \leq j \leq n.$$

The following assertions hold true.

(i) The basic solution of the tableau (5.6) *is an extreme point of S.*
(ii) The basic solution of the tableau (5.7) *is an extreme point of S.*
(iii) The basic solution of all iterative pivots of the Tableau (5.6) *is an extreme point of S.*

Proof. For (i), the basic solution corresponds to the origin. For (ii), the effect of on a nonzero pivot entry $0 \neq a_{ij}$ corresponds to the feasible region's extreme point that has $x_i = b'_j$ and all other x_k, for $k = 1, \dots, \hat{i}, \dots, n$, equal to zero. Similar considerations apply to all subsequent pivots. This establishes item (iii). □

Proof of Simplex Method for basic feasible maximization tableau. Recall that an optimal solution is achieved at an extreme point of the feasible region (see Lemma 3.25). If the basic solution of the given basic feasible solution of the given basic feasible maximization Tucker Tableau is not optimal, then some objective function coefficient c_j must be positive. If for this j, it holds true that $a_{1j}, \dots, a_{mj} \leq 0$, then it is clear that the feasible region S is not bounded. On the other hand, if some $a_{ij} > 0$, then the method proceeds and results in pivoting with respect to a suitable (uniquely defined) nonzero pivot entry a_{ij}.

Under this pivot transform, the basic solution of the resulting tableau corresponds to an extreme point of the feasible region by Lemma 5.10. Further, the objective function does not decrease. (This follows directly by examining the effect of the Pivot Transform Method on the relabelled tableau entry $d = a_{m+1n+1}$.) Finally, by construction only finitely many iterations of Steps (i)–(v) are possible. Thus the method terminates with a tableau which has basic solution corresponding to an optimal solution of the given canonical form maximization linear programming problem. □

Proof of Simplex Method for arbitrary maximization tableau. Considering Steps (i) through (vi) of the Simplex Method, we need only to note that the tableau that is produced by the Pivot Transform Method, that is performed in Steps (iv) or (v) of the Simplex Method, as the case may be, has $b_m > 0$, in the case of Step (iv) and $b_\ell > 0$ in the case of Step (v).

The conclusion is thus that the method iterates Steps (i)–(v) a finite number of times before the resulting maximization Tucker Tableau is a basic feasible Tucker Tableau. In particular, after a finite number of steps the method reaches Step (vi) and the method determines, after an application of the Simplex Method for basic feasible maximization Tableau, an optimal solution. □

5.6 Simplex Method for Dual Tableau and Fundamental Theorem of Linear Programming

Here, we describe a Simplex Method for dual Tucker Tableau (see Section 5.6.2). The key technical step is a Pivot Transform Method for dual Tucker Tableau (see Section 5.6.1).

As an application of this framework, see Section 5.6.3, we use it to prove the Fundamental Theorem of linear programming (see Theorem 5.11). Prior to describing the Simplex Method for dual Tucker Tableau, and to provide additional motivation for it we describe the setting for the Fundamental Theorem of Linear Programming (Theorem 5.11).

Indeed, the setting for the Fundamental Theorem of Linear Programming is that of a *dual Tucker Tableau*

$$\begin{bmatrix} & x_1 & \cdots & x_n & & \\ y_1 & a_{11} & \cdots & a_{1n} & b_1 & x_{n+1} \\ \vdots & \vdots & \vdots & \vdots & \vdots & \vdots \\ y_m & a_{m1} & \cdots & a_{mn} & b_m & x_{n+m} \\ & c_1 & \cdots & c_n & d & \\ & y_{m+1} & \cdots & y_{m+n} & & \end{bmatrix} = \begin{bmatrix} & x_1 & \cdots & x_n & & \\ y_1 & a_{11} & \cdots & a_{1n} & a_{1n+1} & x_{n+1} \\ \vdots & \vdots & \vdots & \vdots & \vdots & \vdots \\ y_m & a_{m1} & \cdots & a_{mn} & a_{mn+1} & x_{n+m} \\ & a_{m+11} & \cdots & a_{m+1n} & a_{m+1n+1} & \\ & y_{m+1} & \cdots & y_{m+n} & & \end{bmatrix} \tag{5.8}$$

that is associated to a *dual pair* of canonical linear programming problems

- Maximize the affine linear objective function

$$f(x_1, \ldots, x_n) = c_1x_1 + \cdots + c_nx_n - d$$

- Subject to the constraints

$$\begin{matrix} a_{11}x_1 & + \ldots + & a_{1n}x_n & \le b_1 \\ \vdots & \vdots & \vdots & \vdots \\ a_{m1}x_1 & + \ldots + & a_{mn}x_n & \le b_m \end{matrix}$$

  and

$$x_1, \ldots, x_n \ge 0$$

respectively

- Minimize the affine linear objective function

$$g(y_1, \ldots, y_m) = b_1y_1 + \cdots + b_my_m - d$$

- Subject to the constraints

$$\begin{matrix} a_{11}y_1 & + \ldots + & a_{m1}y_m & \ge c_1 \\ \vdots & \vdots & \vdots & \vdots \\ a_{1n}y_1 & + \ldots + & a_{mn}y_m & \ge c_n \end{matrix}$$

  and

$$y_1, \ldots, y_m \ge 0.$$

Within this context, the Fundamental Theorem of Linear Programming can be formulated in the following way.

Theorem 5.11. (Fundamental Theorem of Linear Programming). *Consider the dual pair of canonical maximization/minimization linear programming problems that are associated to a dual Tucker Tableau* (5.8). *Then exactly one of the following mutually exclusive possibilities holds true.*

(i) Both problems have optimal solutions; for these solutions the value of the objective functions are equal. Further, an optimal solution can be found as a basic solution of a dual Tucker Tableau obtained from the given dual Tucker Tableau after applying finitely many iterations of the Simplex Method for dual Tucker Tableau.

(ii) The given canonical maximization linear programming problem is unbounded, whereas the dual canonical minimization linear programming problem is infeasible.

(iii) The given canonical minimization linear programming problem is unbounded, whereas the dual canonical maximization linear programming problem is infeasible.

(iv) Both of the given dual canonical maximization/minimization linear programming problems are infeasible.

We prove Theorem 5.11 in Section 5.6.3. Our approach is based on an extension to the Simplex Method. This is the Simplex Method for dual Tucker Tableau. This is discussed in Section 5.6.2.

5.6.1 Pivot Transform Method for Dual Tucker Tableau

Here, we describe the *Pivot Transform Method* for dual Tucker Tableau.

Pivot Transform Method for dual Tucker Tableau

Given a *dual Tucker Tableau*

$$\begin{bmatrix} & x_1 & \dots & x_n & & \\ y_1 & a_{11} & \dots & a_{1n} & b_1 & x_{n+1} \\ \vdots & \vdots & \vdots & \vdots & \vdots & \vdots \\ y_m & a_{m1} & \dots & a_{mn} & b_m & x_{n+m} \\ & c_1 & \dots & c_n & d & \\ & y_{m+1} & \dots & y_{m+n} & & \end{bmatrix} = \begin{bmatrix} & x_1 & \dots & x_n & & \\ y_1 & a_{11} & \dots & a_{1n} & a_{1n+1} & x_{n+1} \\ \vdots & \vdots & \vdots & \vdots & \vdots & \vdots \\ y_m & a_{m1} & \dots & a_{mn} & a_{mn+1} & x_{n+m} \\ & a_{m+11} & \dots & a_{m+1n} & a_{m+1n+1} & \\ & y_{m+1} & \dots & y_{m+n} & & \end{bmatrix} \tag{5.9}$$

by *pivot transform* with respect to a nonzero *pivot entry*

$$0 \neq p^* = p = a_{ij} \text{ for some } 1 \leq i \leq m \text{ and } 1 \leq j \leq n$$

we mean the dual Tableau

$$\begin{bmatrix} & x_1 & \dots & x_n & & \\ y_1 & a'_{11} & \dots & a'_{1n} & b'_1 & x_{n+1} \\ \vdots & \vdots & \vdots & \vdots & \vdots & \vdots \\ y_m & a'_{m1} & \dots & a'_{mn} & b'_m & x_{n+m} \\ & c'_1 & \dots & c'_n & d' & \\ & y_{m+1} & \dots & y_{m+n} & & \end{bmatrix} = \begin{bmatrix} & x_1 & \dots & x_n & & \\ y_1 & a'_{11} & \dots & a'_{1n} & a'_{1n+1} & x_{n+1} \\ \vdots & \vdots & \vdots & \vdots & \vdots & \vdots \\ y_m & a'_{m1} & \dots & a'_{mn} & a'_{mn+1} & x_{n+m} \\ & a'_{m+11} & \dots & a'_{m+1n} & a'_{m+1n+1} & \\ & y_{m+1} & \dots & y_{m+n} & & \end{bmatrix} \tag{5.10}$$

where a'_{ij} for $i = 1, \dots, m+1$ and $j = 1, \dots, n+1$ are defined as in (5.5).

The output of the Pivot Transform Method for dual Tucker Tableau is a dual Tucker Tableau which has entries defined as in (5.5).

5.6.2 *Simplex Method for Dual Tucker Tableau*

In essence, the Simplex Method for dual Tucker Tableau proceeds by applying the Simplex Method to the corresponding maximization tableau and by performing the same sequence of steps, using the pivots for the dual tableau as described in Section 5.6.1, to the dual Tucker Tableau as appropriate.

Simplex Method for dual Tucker Tableau

Given a dual Tucker Tableau (5.8), apply the same steps as the Simplex Method to the corresponding maximization tableau, as described in Section 5.3.3, while using the pivots for the dual tableau as described in Section 5.6.1, to the dual Tucker Tableau as appropriate.

5.6.3 *Proof of the Fundamental Theorem of Linear Programming*

Our proof of the Fundamental Theorem of Linear Programming (Theorem 5.11) is based on the Simplex Method and observing that the sequence of steps applied to the corresponding maximization tableau simultaneously yield the mutually exclusive outcomes that arise in the conclusion of Theorem 5.11.

Proposition 5.12. *A canonical maximization linear programming problem has an optimal solution* $\mathbf{x}^*$ *if and only if the dual canonical maximization linear programming problem has an optimal solution* $\mathbf{y}^*$*. Furthermore, the respective objective functions have equal value at these respective optimal solutions. Especially*

$$f(\mathbf{x}^*) = g(\mathbf{y}^*).$$

Proof. Assume that the given canonical maximization linear programming problem is not infeasible nor unbounded. Then the Simplex Method applied to the maximization tableau terminates in a basic optimal solution for the given canonical maximization linear programming problem. Apply the same sequence of pivots to the corresponding dual Tucker Tableau. The basic solution of the corresponding canonical minimization is also optimal. For example, this follows by considering the associated negative transpose maximization Tucker Tableau. The converse implication follows similarly by arguing dually. □

Proof of Theorem 5.11. Item (i) follows from Proposition 5.12.

For item (ii), if a canonical maximization linear programming problem is unbounded, then the dual canonical linear programming problem is readily seen to be infeasible. For instance this follows upon inspection of the negative transpose of the given minimization Tucker Tableau.

Similarly, to establish items (iii) and (iv), if a canonical maximization linear programming problem is infeasible, then the dual canonical minimization linear programming problem does not have an optimal solution. Thus, the corresponding canonical minimization linear programming problem is either infeasible or unbounded. This yields, respectively, the conclusions of items (iv) and (iii). □

We conclude this section by proving Theorem 4.7 as an application of the Simplex Method.

Proof of Theorem 4.7. Simply observe that the conclusions of Theorem 4.7 follow as a special case of the conclusions of Theorem 5.11. □

5.7 Pivot Transform and Complementary Slackness for Dual Tableau

Here we revisit the idea of dual linear programming problems and entertain the question as to the extent to which such dual pairs of linear programming problems can be solved simultaneously via a common sequence of pivot transformations. The operation of pivot transformations for dual Tucker Tableau is formulated, in an evident way, by extension of the operation of pivoting applied to maximization Tucker Tableau.

These considerations are illustrated in Example 5.13.

Example 5.13. Consider the dual tableau

$$\begin{bmatrix} & x_1 & x_2 & & \\ y_1 & 1 & 2 & 20 & x_3 \\ y_2 & 2 & 2 & 30 & x_4 \\ y_3 & 2 & 1 & 25 & x_5 \\ & 200 & 150 & 0 & \\ & y_4 & y_5 & & \end{bmatrix} = \begin{bmatrix} & x_1 & x_2 & & \\ y_1 & a_{11} & a_{12} & a_{13} & x_3 \\ y_2 & a_{21} & a_{22} & a_{23} & x_4 \\ y_3 & a_{31} & a_{32} & a_{33} & x_5 \\ & a_{41} & a_{42} & a_{43} & \\ & y_4 & y_5 & & \end{bmatrix}.$$

Pivoting on $p^* = a_{31} = 2$ yields the tableau

$$\begin{bmatrix} & x_5 & x_2 & & \\ y_1 & -1/2 & 3/2 & 15/2 & x_3 \\ y_2 & -1 & 1 & 5 & x_4 \\ y_4 & 1/2 & 1/2 & 25/2 & x_1 \\ & -100 & 50 & -2500 & \\ & y_3 & y_5 & & \end{bmatrix}.$$

Subsequently, pivoting on $p^* = a_{22} = 1$ then yields the tableau

$$\left[\begin{array}{ccccc} & x_5 & x_3 & & \\ y_1 & 1 & -3/2 & 0 & x_3 \\ y_5 & -1 & 1 & 5 & x_2 \\ y_4 & 1 & -1/2 & 10 & x_1 \\ & -50 & -50 & -2750 & \\ & y_3 & y_2 & & \end{array}\right].$$

It turns out that this last tableau simultaneously solves the corresponding maximum and minimum dual linear programming problems.

Indeed,

$$\max f = f(x_5 = 0, x_3 = 0) = f(x_1 = 10, x_2 = 5) = 2750$$

whereas

$$\min g = g(y_1 = 0, y_5 = 0, y_4 = 0) = g(y_1 = 0, y_2 = 50, y_3 = 50) = 2750.$$

Example 5.14. Starting with the dual tableau

$$\left[\begin{array}{ccccc} & x_1 & x_2 & & \\ y_1 & 20 & 25 & 300 & x_3 \\ y_2 & 40 & 20 & 500 & x_4 \\ & 1000 & 800 & 0 & \\ & y_3 & y_4 & & \end{array}\right]$$

pivoting on $p^* = a_{21} = 40$ and then pivoting on $p^* = a_{12} = 15$ yields the respective tableaus

$$\left[\begin{array}{ccccc} & x_4 & x_2 & & \\ y_1 & -1/2 & 15 & 50 & x_3 \\ y_3 & 1/40 & 1/2 & 25/2 & x_1 \\ & -25 & 300 & -12500 & \\ & y_2 & y_4 & & \end{array}\right]$$

and

$$\left[\begin{array}{ccccc} & x_4 & x_3 & & \\ y_4 & -1/30 & 1/15 & 10/3 & x_2 \\ y_3 & 1/24 & -1/30 & 65/6 & x_1 \\ & -15 & -20 & -13500 & \\ & y_2 & y_1 & & \end{array}\right].$$

The basic solution of each programming problem in the final tableau is optimal. Especially

$$\max f = f(x_4 = 0, x_3 = 0) = f(x_1 = 65/6, x_2 = 10/3) = 13500$$

and

$$\min g = g(y_3 = 0, y_4 = 0) = g(y_1 = 20, y_2 = 15) = 13500.$$

Exercises

5.1 Consider the maximization linear programming problem

- Maximize the linear objective function

$$f(x_1, x_2) = 20x_1 + 10x_2$$

- Subject to the constraints

$$\begin{aligned} x_1 + 2x_2 &\le 40 \\ x_1 + 2x_2 &\le 60 \end{aligned}$$

  and

$$x_1, x_2 \ge 0.$$

(i) Determine the corresponding slack linear programming problem.
(ii) Determine the corresponding maximization tableau.
(iii) Use the Simplex Method to determine an optimal solution.

5.2 Describe step by step the Simplex Method applied to the linear programming problem

- Maximize the linear objective function

$$f(x_1, x_2) = 2x_1 + 3x_2$$

- Subject to the constraints

$$\begin{aligned} -3x_1 + x_2 &\le 1 \\ 4x_1 + 2x_2 &\le 20 \\ 4x_1 - x_2 &\le 10 \\ -x_1 + 2x_2 &\le 5 \end{aligned}$$

  and

$$x_1, x_2 \ge 0.$$

5.3 Describe step by step the Simplex Method applied to the linear programming problem

- Maximize the linear objective function

$$f(x_1, x_2, x_3) = 2x_1 + 4x_2 + 3x_3$$

- Subject to the constraints

$$\begin{aligned} 3x_1 + \ x_2 \ + 2x_3 &\leq 40 \\ 2x_1 + \ x_2 \ + 2x_3 &\leq 40 \\ x_1 \ + 2x_2 + 2x_3 &\leq 80 \end{aligned}$$

and

$$x_1, x_2, x_3 \geq 0.$$

5.4 Consider the minimization linear programming problem

- Minimize the linear objective function

$$g(y_1, y_2) = 10y_1 + 6y_2 + 8y_2$$

- Subject to the constraints

$$\begin{aligned} y_1 \ + \ y_2 \ + 2y_3 &\geq 2 \\ 5y_1 + 3y_2 + 2y_3 &\geq 1 \end{aligned}$$

and

$$y_1, y_2, y_3 \geq 0.$$

Use the Simplex Method applied to the corresponding negative transpose tableau to determine an optimal solution.

5.5 Consider the Tucker Tableau

$$\begin{bmatrix} x_1 & x_2 & & \\ -1 & -2 & -3 & x_3 \\ 1 & 1 & 3 & x_4 \\ 1 & 1 & 2 & x_5 \\ -2 & 4 & 0 & \end{bmatrix}.$$

Use the Simplex Method to determine an optimal solution to the canonical maximization linear programming problem.

5.6 Use the technique of negative transpose to find an optimal solution of the canonical minimization linear programming problem that is associated to the minimization Tucker Tableau

$$\begin{bmatrix} y_1 & 20 & 25 & 300 \\ y_2 & 40 & 20 & 500 \\ & 1000 & 800 & 0 \\ & y_3 & y_4 & \end{bmatrix}.$$

Chapter 6
Using Computer Software to Solve Linear Programming Problems

Overview

- Selected Python scripts for the Simplex Method.
- Solving linear programming problems using the Python package scipy.optimize.
- Solving linear programming problems using the Algebraic Modelling Language Pyomo.

6.1 Selected Python Scripts for the Simplex Method

In this section, we describe a collection of Python scripts which allow for a direct implementation of the Simplex Method, as described in Chapter 5. These scripts, see Sections 6.1.1, 6.1.2 and 6.1.3, can be used to reproduce each of the Simplex Method related examples that are described there. They illustrate the most basic fundamental concepts that underlie more robust larger scale implementations of linear programming techniques.

Two such larger scale solvers, include the HiGHS and glpk solvers. More information about these solvers can be found on the respective websites

https://highs.dev

and

https://pypi.org/project/glpk/.

These solvers give an implementation of the dual Simplex Method. Their most basic features are illustrated in Sections 6.2, and 6.3, using, respectively, the scipy.optimize and Pyomo platforms. The glpk solver also makes use of interior point methods for solving linear programming problems.

N. Grieve, *An Introduction to Mathematical Programming and Network Science*,
Springer Undergraduate Texts in Mathematics and Technology,
https://doi.org/10.1007/978-3-032-13330-4_6

More details about these platforms can be found on the respective websites

https://docs.scipy.org/doc/scipy/reference/optimize.html

and

https://www.pyomo.org.

6.1.1 Python Scripts for the Pivot Transform Method

The following scripts are based on a direct realization of a given Tucker Tableau as a two-dimensional array. Given such a description of the tableau, the scripts that follow allow for a direct implementation of the Pivot Transform Method (as described in Section 5.3.1).

For our purposes, the most direct way to represent a Tucker Tableau is to represent it directly as a double index array and then to keep track of the row/column shift, as appropriate when pivoting along a particular entry. This approach is utilized in what follows.

What follows below, in the Python scripts for the Pivot Transform Method, are three methods. The first returns the size of a tableau. The second returns the particular tableau entry upon pivoting the tableau with respect to a particular tableau entry. The third method returns the entire tableau upon pivoting with respect to a particular tableau entry. Together these three methods comprise the Pivot Transform Method.

Python scripts for the Pivot Transform Method

```
# A method which returns the size of a given tableau.

def tableau_size(tableau):
    return [len(tableau),len(tableau[0])]

# A method which returns the Tucker Tableau entries
# under the operation of pivoting
# with respect to a given pivot entry.

def tableau_pivot_function(tableau, pivot_entry_index,entries):
    t = tableau
    tab_size = tableau_size(t)
    m = tab_size[0]
    n = tab_size[1]
    pivot_row_index = pivot_entry_index[0]
    pivot_col_index = pivot_entry_index[1]
    pivot_entry = tableau[pivot_row_index][pivot_col_index]
    i = entries[0]
    j = entries[1]
    tableau_entry = t[i][j]
    if i ==0 and j == pivot_col_index:
                return tableau[pivot_row_index][n-1]
    if i ==0 and j != pivot_col_index:
        return tableau[i][j]
```

```
    if i == pivot_row_index and j == n-1:
                return t[0][pivot_col_index]
    if i != pivot_row_index and j == n-1:
        return t[i][j]
    if i == pivot_row_index and j == pivot_col_index:
                return 1/(pivot_entry)
    if i == pivot_row_index and j != pivot_col_index and j != n-1:
                return tableau_entry / pivot_entry
    if i != pivot_row_index and j == pivot_col_index:
                return - tableau_entry / pivot_entry
    if i !=0 and (i != pivot_row_index and j != pivot_col_index) and j!= n -1:
                q = t[pivot_row_index][j]
                r = t[i][pivot_col_index]
                return tableau_entry - q*r/pivot_entry
    if i == m-1 and j == n-1:
        return t[i][j]

#  A method which returns the Tucker Tableau that
# is obtained via pivoting
# with respect to a given pivot entry index.

def pivot_transform(tableau, pivot_entry_index):
    t = tableau
    tab_size = tableau_size(t)
    m = tab_size[0]
    n = tab_size[1]
    pivot_tableau = []
    for i in range(m):
        row_i = []
        for j in range(n):
            row_i.append(tableau_pivot_function(tableau, pivot_entry_index,[i,j]))
        pivot_tableau.append(row_i)
    return  pivot_tableau
```

The above scripts are illustrated in Examples 6.1 below.

Example 6.1. As in Example 5.7, consider the Tucker Tableau

$$\begin{bmatrix} x_1 & x_2 & & \\ -1 & -2 & -3 & x_3 \\ 1 & 1 & 3 & x_4 \\ 1 & 1 & 2 & x_5 \\ -2 & 4 & 0 & \end{bmatrix}.$$

The above methods, to implement the Pivot Transform Method (and other methods that follow) are saved in an auxiliary file Simplex_Tableau.py. After importing this file, the respective pivots along tableau entries $(3, 1)$, $(1, 2)$, $(3, 1)$ and $(3, 2)$ are achieved as described below.

Python scripts for Example 6.1

```
>>> import Simplex_Tableau as SimpTab
>>> A1 = [[1,2," ", " "],[-1,-2,-3,3],[1,1,3,4],[1,1,2,5],[-2,4,0," "]]
>>> print(A1)
[[1, 2, ' ', ' '], [-1, -2, -3, 3], [1, 1, 3, 4], [1, 1, 2, 5], [-2, 4, 0, ' ']]
>>> SimpTab.tableau_size(A1)
[5, 4]
>>> A2 = SimpTab.pivot_transform(A1,[3,0])
>>> print(A2)
[[5, 2, ' ', ' '], [1.0, -1.0, -1.0, 3], [-1.0, 0.0, 1.0, 4], [1.0, 1.0, 2.0, 1],
[2.0, 6.0, 4.0, ' ']]
>>> A3 = SimpTab.pivot_transform(A2,[1,1])
>>> print(A3)
[[5, 3, ' ', ' '], [-1.0, -1.0, 1.0, 2], [-1.0, 0.0, 1.0, 4], [2.0, 1.0, 1.0, 1],
[8.0, 6.0, -2.0, ' ']]
>>> A4 = SimpTab.pivot_transform(A3,[3,0])
>>> print(A4)
[[1, 3, ' ', ' '], [0.5, -0.5, 1.5, 2], [0.5, 0.5, 1.5, 4], [0.5, 0.5, 0.5, 5],
[-4.0, 2.0, -6.0, ' ']]
>>> A5 = SimpTab.pivot_transform(A4,[3,1])
>>> print(A5)
[[1, 5, ' ', ' '], [1.0, 1.0, 2.0, 2], [0.0, -1.0, 1.0, 4], [1.0, 2.0, 1.0, 3],
[-6.0, -4.0, -8.0, ' ']]
```

In the above scripts, observe that the resulting tableau, represented as arrays, correspond to those that are produced in Example 5.7. Note also that due to the column indexing conventions, pivoting along tableau entry (i, j) corresponds to pivoting along array entry $[i, j - 1]$.

6.1.2 Python Scripts for Basic Feasible Tableau

We now give a collection of scripts which pertain to basic feasible tableau. These are based on the methods from Section 5.3.2. Combined with the implementation of the Pivot Transform Method, as implemented in Section 6.1.1, we obtain a direct implementation of the Simplex Method. This is described in Section 6.1.3.

What follows below in the Python scripts for basic feasible tableau is a collection of four methods. The first method checks if a given tableau is basic feasible. The second checks if the basic solution of a given basic feasible tableau is optimal. Before doing this, the method checks to see if the given input is actually a basic feasible tableau. The third method finds a suitable pivot entry for the case of a non-optimal basic feasible tableau to pivot along so as to make progress towards obtaining an optimal basic feasible tableau. First, the method checks to see if the given input is actually a basic feasible tableau. The fourth method returns, for the case of a non-basic feasible tableau, a suitable entry to pivot along so as to make progress towards achieving a basic feasible tableau. The first step is to check to see if the given tableau is optimal.

Python scripts for basic feasible tableau

```
# A method to check if a tableau is basic feasible.

def is_basic_feasible(tableau):
    t = tableau
    tab_size = tableau_size(t)
    m = tab_size[0]
    n = tab_size[1]
    L = [t[i+1][n-2] for i in range(m-2) if t[i+1][n-2]<0]
    if len(L) > 0:
        return False
    else:
        return True

# A method to check if a
# basic feasible tableau is optimal.

def is_basic_feasible_optimal(tableau):
    t = tableau
    tab_size = tableau_size(t)
    m = tab_size[0]
    n = tab_size[1]
    if is_basic_feasible(tableau) == False:
        return "Tableau is not basic feasible"
    else:
        L = [t[m-1][j] for j in range(n-2) if t[m-1][j]>0]
        if len(L) > 0:
            return False
        else:
            return True

# A method to find suitable pivot
# row and column index for the case of
# not optimal basic feasible tableau.

def bf_choose_pivot_index(tableau):
    t = tableau
    tab_size = tableau_size(t)
    m = tab_size[0]
    n = tab_size[1]
    if is_basic_feasible(t) == False:
        return "Tableau is not basic feasible"
    if is_basic_feasible_optimal(t) == True:
        return "Tableau is optimal"
    else:
        L = [k for k in range(n-2) if t[m-1][k]>0]
        j = min(L)
        a = [t[i+1][j] for i in range(m-2) if t[i+1][j]>0]
        if len(a) == 0:
            return "Tableau is unbounded"
        else:
            M = min( [(t[i+1][n-2])/(t[i+1][j]) for i in range(m-2) if t[i+1][j]>0])
            P = [i for i in range(m-2) if t[i+1][j] > 0]
            i = min([i+1 for i in P if ((t[i+1][n-2])/(t[i+1][j]))==M])
            return([i,j])

# A method which has output a pivot entry
# that will assist with pivoting towards
# a basic feasible maximization tableau
# into a basic feasible tableau.

def pivot_towards_bf(tableau):
    t = tableau
    if is_basic_feasible(t) == True:
```

```
        return "The given Tableau is a basic feasible Tableau"
    tab_size = tableau_size(t)
    m = tab_size[0]
    n = tab_size[1]
    i = max([k for k in range(1,m-1) if t[k][n-2] < 0])
    a = [t[i][j] for j in range(n-2) if t[i][j] < 0]
    if len(a) == 0:
        return "The given Tableau is not feasible"
    j = min([j for j in range(n-2) if t[i][j]<0])
    if i == m-2:
        return [i,j]
    else:
        L = [(t[k][n-2])/(t[k][j]) for k in range(i+1,m-1) if t[k][j]>0]
        L.append(t[i][n-2]/t[i][j])
        M = min(L)
        if M == t[i][n-2]/t[i][j]:
            return [i,j]
        else:
            l = min([k for k in range(i+1,m-1) if t[k][j]>0
                     and M == (t[k][n-2])/(t[k][j])])
```

These scripts are illustrated in Section 6.1.3, see Example 6.2, which also serves to illustrate a direct Python illustration of the Simplex Method.

6.1.3 A Direct Python Implementation of the Simplex Method

In this section, we illustrate how together the scripts given in Sections 6.1.1 and 6.1.2 allow for a direct implementation of the Simplex Method. (See Example 6.2.)

Example 6.2. Let us illustrate Example 6.1 from the view point of the scripts given in Sections 6.1.1 and 6.1.2. We start as input the Tucker Tableau that is given there.

Python scripts for Example 6.2

```
>>> import Simplex_Tableau as SimpTab
>>> A1 = [[1,2," ", " "],[-1,-2,-3,3],[1,1,3,4],[1,1,2,5],[-2,4,0," "]]
>>> SimpTab.is_basic_feasible(A1)
False
>>> SimpTab.pivot_towards_bf(A1)
[3, 0]
>>> A2 = SimpTab.pivot_transform(A1,[3,0])
>>> SimpTab.pivot_towards_bf(A2)
[1, 1]
>>> A3 = SimpTab.pivot_transform(A2,[1,1])
>>> SimpTab.pivot_towards_bf(A3)
'The given Tableau is a basic feasible Tableau'
>>> SimpTab.is_basic_feasible_optimal(A3)
False
>>> SimpTab.bf_choose_pivot_index(A3)
[3, 0]
>>> A4 = SimpTab.pivot_transform(A3,[3,0])
>>> SimpTab.is_basic_feasible_optimal(A4)
False
>>> SimpTab.bf_choose_pivot_index(A4)
```

```
[3, 1]
>>> A5 = SimpTab.pivot_transform(A4,[3,1])
>>> SimpTab.is_basic_feasible_optimal(A5)
True
>>> print(A5)
[[1, 5, ' ', ' '], [1.0, 1.0, 2.0, 2], [0.0, -1.0, 1.0, 4], [1.0, 2.0, 1.0, 3],
[-6.0, -4.0, -8.0, ' ']]
```

6.2 Solving Linear Programming Problems Using the Python Package scipy.optimize

In Examples 6.3, 6.4 and 6.5 below, we indicate how to use the most basic linear programming features of the Python package scipy.optimize. More details about the scipy.optimize package can be found at its documentation website

https://docs.scipy.org/doc/scipy/reference/optimize.html.

Here, we illustrate how to solve linear programming problems using the scipy.optimize implementation of the HiGHS implementation of the dual Simplex Method. More details about the HiGHS project can be found at

https://highs.dev.

Example 6.3. Consider the two variable maximization linear programming problem

- Maximize the linear objective function

$$Z = 10x + 20y$$

- Subject to the constraints

$$\begin{aligned} -x + 2y &\leq 15 \\ x + y &\leq 12 \\ 5x + 3y &\leq 45 \end{aligned}$$

  and

$$x, y \geq 0.$$

The following Python session solves the given maximization linear programming problem indicating that the maximal value of the objective function is $Z(x, y) = 210$ and that it is achieved when $x = 3$ and $y = 9$.

Python scipy.optimize scripts for Example 6.3

```
>>> from scipy.optimize import linprog
>>> obj_fun = [-10,-20] # min - Z = - 10 x - 20 y
... lhs_ineq = [[-1,2],# -x + 2y
...             [1,1], # x + y
...             [5,3]] # 5x + 3y
... rhs_ineq = [15, # <= 15
...             12, #  <= 12
...             45] # <= 45
>>> opt_soltn = linprog(c=obj_fun, A_ub=lhs_ineq, b_ub=rhs_ineq,method="highs")
>>> print(opt_soltn)
        message: Optimization terminated successfully. (HiGHS Status 7: Optimal)
        success: True
         status: 0
            fun: -210.00000000000003
              x: [ 3.000e+00  9.000e+00]
            nit: 3
          lower:  residual: [ 3.000e+00  9.000e+00]
                marginals: [ 0.000e+00  0.000e+00]
          upper:  residual: [       inf        inf]
                marginals: [ 0.000e+00  0.000e+00]
          eqlin:  residual: []
                marginals: []
        ineqlin:  residual: [ 0.000e+00  0.000e+00  3.000e+00]
                marginals: [-3.333e+00 -1.333e+01 -0.000e+00]
 mip_node_count: 0
 mip_dual_bound: 0.0
        mip_gap: 0.0
```

In using the scipy.optimize package, we have minimized the linear programming problem that is associated to the objective function $-Z$. The reason is that the default implementation of the methods for linear programming problems are configured to solve minimization, as opposed to maximization, linear programming problems. We refer to the scipy.optimize documentation for more details about the various output parameters.

For the purpose of the given maximization optimization problem at hand, the output line fun: indicates that $-Z$ has minimum value -210 when $x = 3$ and $y = 9$. The conclusion is that Z has maximum value

$$Z(x, y) = 210$$

when $x = 3$ and $y = 9$.

The given programming problem's feasible region is pictured in Figure 6.1.

Example 6.4. Consider the two variable minimization linear programming problem

- Minimize the linear objective function

$$Z = 15x + 20y$$

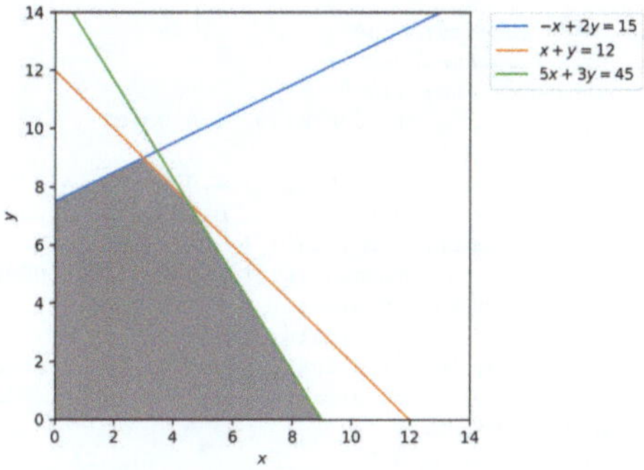

Fig. 6.1 The feasible region for Example 6.3

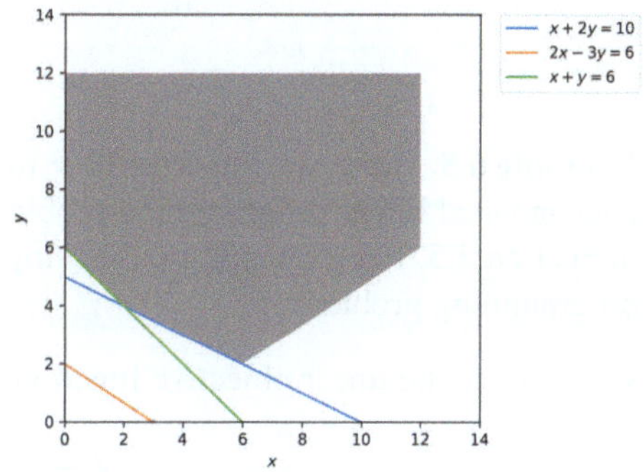

Fig. 6.2 The feasible region for Example 6.4

- Subject to the constraints

$$\begin{aligned} x + 2y &\geq 10 \\ 2x - 3y &\leq 6 \\ x + y &\geq 6 \end{aligned}$$

  and

$$x, y \geq 0.$$

The feasible region is pictured in Figure 6.2.

The following script uses the scipy.optimize package to solve the given minimization linear programming problem. The output indicates that the minimum value of the objective function is 110 and that it is achieved when $x = 2$ and $y = 4$.

Python scipy.optimize scripts for Example 6.4

```
>>> obj_fun = [15,20] # min  Z = 15 x + 20 y
... lhs_ineq = [[-1,-2],# x + 2y
...             [2,-3], # 2x + 3y
...             [-1,-1]] # x + y
... rhs_ineq = [-10, # >= 10
...             6, #  <= 6
...             -6] # >= 6
... opt_soltn = linprog(c=obj_fun, A_ub=lhs_ineq, b_ub=rhs_ineq,method="highs")
... print(opt_soltn)
...
        message: Optimization terminated successfully. (HiGHS Status 7: Optimal)
```

```
        success: True
         status: 0
            fun: 110.0
              x: [ 2.000e+00  4.000e+00]
            nit: 2
          lower:  residual: [ 2.000e+00  4.000e+00]
                 marginals: [ 0.000e+00  0.000e+00]
          upper:  residual: [       inf        inf]
                 marginals: [ 0.000e+00  0.000e+00]
          eqlin:  residual: []
                 marginals: []
        ineqlin:  residual: [ 0.000e+00  1.400e+01  0.000e+00]
                 marginals: [-5.000e+00 -0.000e+00 -1.000e+01]
 mip_node_count: 0
 mip_dual_bound: 0.0
        mip_gap: 0.0
```

Example 6.5. Here we illustrate how to use scipy.optimize to solve more general noncanonical linear programming problems—for instance, those that are discussed in Section 3.5. For example, to use scipy.optimize to solve the minimization linear programming problem

- Minimize the linear objective function

$$Z = x + 2y + 3z$$

- Subject to the constraints

$$\begin{aligned} 3x \qquad + 4z &\leq 5 \\ 5x + y + 6z &= 7 \\ 8x \qquad + 9z &\leq 2 \end{aligned}$$

  and

$$x, y, z \geq 0$$

there is no need to re-express it as a suitable canonical form minimization linear programming problem.

Instead, we may proceed directly as illustrated in the following way.

Python scipy.optimize scripts for Example 6.5

```
>>> obj_func = [1,2,3] # min  Z =  x + 2 y + 3 z
... lhs_ineq = [[3,0,4],# 3x + 4z
...             [8,0,9]] # 8x + 9z
... rhs_ineq = [5, # 5
...             2] # 2
... lhs_eq = [[5,1,6]] # 5x + y + 6z
... rhs_eq = [7] # 7
... opt_soltn = linprog(c=obj_func, A_ub=lhs_ineq, b_ub=rhs_ineq, A_eq = lhs_eq,
... b_eq = rhs_eq, method="highs")
... print(opt_soltn)
...
```

```
        message: Optimization terminated successfully. (HiGHS Status 7: Optimal)
        success: True
         status: 0
            fun: 11.75
              x: [ 2.500e-01  5.750e+00  0.000e+00]
            nit: 2
          lower:  residual: [ 2.500e-01  5.750e+00  0.000e+00]
                 marginals: [ 0.000e+00  0.000e+00  1.125e+00]
          upper:  residual: [       inf        inf        inf]
                 marginals: [ 0.000e+00  0.000e+00  0.000e+00]
          eqlin:  residual: [ 0.000e+00]
                 marginals: [ 2.000e+00]
        ineqlin:  residual: [ 4.250e+00  0.000e+00]
                 marginals: [-0.000e+00 -1.125e+00]
mip_node_count: 0
mip_dual_bound: 0.0
       mip_gap: 0.0
```

The conclusion is that the optimal solution

$$Z(x, y, z) = 11.75$$

is achieved when $x_1 = .25$, $y = 5.75$ and $z = 0$.

6.3 Solving Linear Programming Problems Using the Algebraic Modelling Language Pyomo

In this section, we illustrate how to use the most basic features of the Algebraic Modelling Language Pyomo to solve linear programming problems. More details about the Pyomo project can be found at

https://www.pyomo.org

and in the reference book [5].

The Pyomo platform allows for solving mathematical programming problems via a variety of third party solvers. For our purposes here, we use the glpk solver within the Pyomo environment. More information about the glpk solver can be found at

https://pypi.org/project/glpk/.

Example 6.6. In this example, we illustrate how to use Pyomo to solve the maximization linear programming problem

- Maximize the linear objective function

$$f(x_1, x_2, x_3) = 2x_1 + 3x_2 + 4x_3$$

- Subject to the constraints

$$\begin{aligned} x_1 + x_2 + x_3 &\leq 1 \\ x_1 + x_2 + 2x_3 &= 2 \\ 3x_1 + 2x_2 + x_3 &\geq 4 \end{aligned}$$

  and

$$x_1, x_2, x_3 \geq 0.$$

In the following scripts, we use Pyomo to create an abstract modelling environment which encodes the given maximization linear programming problem. We then use the glpk solver, within the Pyomo environment, to solve for an optimal solution.

Python Pyomo scripts for Example 6.6

```
>>> #import pyomo
... import pyomo.environ as pyo
... # Create a model within
... # the Pyomo environment
... # for the particular programming
... # problem at hand
... model = pyo.ConcreteModel() # initialize the model
... # initialize the decision variables
... model.x_1 = pyo.Var(within=pyo.NonNegativeReals)
... model.x_2 = pyo.Var(within=pyo.NonNegativeReals)
... model.x_3 = pyo.Var(within=pyo.NonNegativeReals)
... # create the objective function
... model.obj = pyo.Objective(\
...     expr= 2*model.x_1 + 3*model.x_2 + 4*model.x_3, sense=pyo.maximize)
... # create the constraints
... model.con1 = pyo.Constraint(\
...     expr= model.x_1 + model.x_2 + model.x_3 <= 1)
... model.con2 = pyo.Constraint(\
...     expr= 3*model.x_1 + 2*model.x_2 + model.x_3 == 2)
... model.con3 = pyo.Constraint(\
...     expr= 3*model.x_1 + 2*model.x_2 + model.x_3<= 4)
...
... # make the solver interface and solve the model
... # using the glpk solver
... solver = pyo.SolverFactory('glpk')
... solver.solve(model) # solve the model
... model.display()  # display the output
...
Model unknown
  Variables:
    x_1 : Size=1, Index=None
        Key  : Lower : Value : Upper : Fixed : Stale : Domain
        None :     0 :   0.5 :  None : False : False : NonNegativeReals
    x_2 : Size=1, Index=None
        Key  : Lower : Value : Upper : Fixed : Stale : Domain
        None :     0 :   0.0 :  None : False : False : NonNegativeReals
    x_3 : Size=1, Index=None
        Key  : Lower : Value : Upper : Fixed : Stale : Domain
        None :     0 :   0.5 :  None : False : False : NonNegativeReals
  Objectives:
    obj : Size=1, Index=None, Active=True
        Key  : Active : Value
        None :   True :   3.0
```

```
  Constraints:
    con1 : Size=1
        Key  : Lower : Body : Upper
        None :  None :  1.0 :   1.0
    con2 : Size=1
        Key  : Lower : Body : Upper
        None :   2.0 :  2.0 :   2.0
    con3 : Size=1
        Key  : Lower : Body : Upper
        None :  None :  2.0 :   4.0
```

The above output indicates that the optimal value of the objective function is

$$f(x_1, x_2, x_3) = 3.$$

This is achieved when $x_1 = 0.5$, $x_2 = 0$ and $x_3 = 0.5$.

Example 6.7. Here we illustrate how to use Pyomo to solve the minimization linear programming problem considered in Example 6.5.

Python Pyomo scripts for Example 6.7

```
>>> #import pyomo
... import pyomo.environ as pyo
... # Create a model within
... # the Pyomo environment
... # for the particular programming
... # problem at hand
... model = pyo.ConcreteModel() # initialize the model
... # initialize the decision variables
... model.x_1 = pyo.Var(within=pyo.NonNegativeReals)
... model.x_2 = pyo.Var(within=pyo.NonNegativeReals)
... model.x_3 = pyo.Var(within=pyo.NonNegativeReals)
... # create the objective function
... model.obj = pyo.Objective(\
...     expr= model.x_1 + 2*model.x_2 + 3*model.x_3, sense=pyo.minimize)
... # create the constraints
... model.con1 = pyo.Constraint(\
...     expr= 3*model.x_1  + 4*model.x_3 <= 5)
... model.con2 = pyo.Constraint(\
...     expr= 5*model.x_1 + model.x_2 + 6*model.x_3 == 7)
... model.con3 = pyo.Constraint(\
...     expr= 8*model.x_1 + 9*model.x_3<= 2)
...
... # make the solver interface and solve the model
... # using the glpk solver
... solver = pyo.SolverFactory('glpk')
... solver.solve(model) # solve the model
... model.display()  # display the output
...
Model unknown
  Variables:
    x_1 : Size=1, Index=None
        Key  : Lower : Value : Upper : Fixed : Stale : Domain
        None :     0 :  0.25 :  None : False : False : NonNegativeReals
    x_2 : Size=1, Index=None
        Key  : Lower : Value : Upper : Fixed : Stale : Domain
        None :     0 :  5.75 :  None : False : False : NonNegativeReals
```

```
    x_3 : Size=1, Index=None
        Key  : Lower : Value : Upper : Fixed : Stale : Domain
        None :     0 :   0.0 :  None : False : False : NonNegativeReals
  Objectives:
    obj : Size=1, Index=None, Active=True
        Key  : Active : Value
        None :   True : 11.75
  Constraints:
    con1 : Size=1
        Key  : Lower : Body : Upper
        None :  None : 0.75 :   5.0
    con2 : Size=1
        Key  : Lower : Body : Upper
        None :   7.0 :  7.0 :   7.0
    con3 : Size=1
        Key  : Lower : Body : Upper
        None :  None :  2.0 :   2.0
```

As illustrated in the above output, the objective function has minimum value

$$Z(x_1, x_2, x_3) = 11.75$$

and this is achieved when $x_1 = 0.25$, $x_2 = 5.75$ and $x_3 = 0$.

6.4 Scripts for Diet Problems, Mixture Blending, Agricultural Planning and Scheduling

Example 6.8. Here, we illustrate how to solve, using scipy.omtimize and Pyomo, respectively, the mixture blending problem considered in Example 2.2.

Python scipy.optimize scripts for Example 2.2

```
>>> from scipy.optimize import linprog
... obj_func = [.5,.25,.9,.25,3.25,1.25]
... lhs_ineq = [[-.16,-.18,-.15,-.16,-.125,-.170],
...             [-13,-2.4,-10,-9,-20,-3],
...              [-50,-150,-400,-26,-24,-27]]
... rhs_ineq = [-.4,-30,-450]
... opt_soltn = linprog(c=obj_func, A_ub=lhs_ineq, b_ub=rhs_ineq,method="highs"\
)
... print(opt_soltn)
...
        message: Optimization terminated successfully. (HiGHS Status 7: Optimal)
        success: True
         status: 0
            fun: 1.2989282385834109
              x: [ 0.000e+00  2.540e+00  0.000e+00  2.656e+00  0.000e+00
                   0.000e+00]
            nit: 2
          lower:  residual: [ 0.000e+00  2.540e+00  0.000e+00  2.656e+00
                              0.000e+00  0.000e+00]
                 marginals: [ 1.229e-01  0.000e+00  1.467e-01  0.000e+00
                              2.738e+00  1.143e+00]
```

```
          upper:  residual: [         inf          inf          inf          inf
                                      inf          inf]
                 marginals: [ 0.000e+00  0.000e+00  0.000e+00  0.000e+00
                              0.000e+00  0.000e+00]
          eqlin:  residual: []
                 marginals: []
        ineqlin:  residual: [ 4.821e-01  0.000e+00  0.000e+00]
                 marginals: [-0.000e+00 -2.408e-02 -1.281e-03]
 mip_node_count: 0
 mip_dual_bound: 0.0
        mip_gap: 0.0
```

Python Pyomo scripts for Example 2.2

```
>>> # We are directly initializing the decision
... # variables, inputing the objective function
... # and constraints. Input efficiencies can
... # be achieved by writing input wrapper scripts.
... import pyomo.environ as pyo
...
... model = pyo.ConcreteModel()
... model.x_1 = pyo.Var(within=pyo.NonNegativeReals)
... model.x_2 = pyo.Var(within=pyo.NonNegativeReals)
... model.x_3 = pyo.Var(within=pyo.NonNegativeReals)
... model.x_4 = pyo.Var(within=pyo.NonNegativeReals)
... model.x_5 = pyo.Var(within=pyo.NonNegativeReals)
... model.x_6 = pyo.Var(within=pyo.NonNegativeReals)
...
... model.obj = pyo.Objective(\
...     expr= .5*model.x_1 +.25*model.x_2
...     +.9*model.x_3 +.25*model.x_4
...     +3.25*model.x_5+1.25*model.x_6,
...     sense=pyo.minimize)
...
... model.Con1 = pyo.Constraint(\
...     expr= .16*model.x_1 + .18*model.x_2
...     + .15*model.x_3 + .16*model.x_4
...     + .125*model.x_5 + .170*model.x_6 >= .4)
... model.Con2 = pyo.Constraint(\
...     expr= 13*model.x_1 + 2.4*model.x_2
...     + 10*model.x_3 + 9*model.x_4
...     +20*model.x_5 + 3*model.x_6 >= 30)
... model.Con3 = pyo.Constraint(\
...     expr= 50*model.x_1 + 150*model.x_2
...     + 400*model.x_3 + 26*model.x_4
...     + 24*model.x_5 + 27*model.x_6 >= 450)
...
... # create the solver interface and solve the model
... solver = pyo.SolverFactory('glpk')
... solver.solve(model)
... model.display()
...
Model unknown

  Variables:
    x_1 : Size=1, Index=None
        Key  : Lower : Value : Upper : Fixed : Stale : Domain
        None :     0 :   0.0 :  None : False : False : NonNegativeReals
```

```
  x_2 : Size=1, Index=None
      Key  : Lower : Value            : Upper : Fixed : Stale : Domain
      None :     0 : 2.53960857409133 :  None : False : False : NonNegativeReals
  x_3 : Size=1, Index=None
      Key  : Lower : Value : Upper : Fixed : Stale : Domain
      None :     0 :   0.0 :  None : False : False : NonNegativeReals
  x_4 : Size=1, Index=None
      Key  : Lower : Value            : Upper : Fixed : Stale : Domain
      None :     0 : 2.65610438024231 :  None : False : False : NonNegativeReals
  x_5 : Size=1, Index=None
      Key  : Lower : Value : Upper : Fixed : Stale : Domain
      None :     0 :   0.0 :  None : False : False : NonNegativeReals
  x_6 : Size=1, Index=None
      Key  : Lower : Value : Upper : Fixed : Stale : Domain
      None :     0 :   0.0 :  None : False : False : NonNegativeReals

Objectives:
  obj : Size=1, Index=None, Active=True
      Key  : Active : Value
      None :   True : 1.29892823858341

Constraints:
  Con1 : Size=1
      Key  : Lower : Body              : Upper
      None :   0.4 : 0.882106244175209 :  None
  Con2 : Size=1
      Key  : Lower : Body               : Upper
      None :  30.0 : 29.999999999999986 :  None
  Con3 : Size=1
      Key  : Lower : Body               : Upper
      None : 450.0 : 449.99999999999955 :  None
```

Example 6.9. Here, we illustrate how to solve, using scipy.omtimize and Pyomo, respectively, the mixture blending problem considered in Example 2.3. In this example, it is helpful to note that the output produced by the glpk solver is different than the output produced by the HiGHS solver. Indeed, the output produced by the glpk solver produces interior points of the feasible region at which the objective function achieved its optimal value.

Python scipy.optimize scripts for Example 2.3

```
>>> from scipy.optimize import linprog
... obj_func = [-64.5,-64.5,-64.5,-64.5,-59.91,-59.91,-59.91,-59.91]
... lhs_ineq = [[-115,-105,-119,-101,0,0,0,0],
...             [0,0,0,0,-115,-105,-119,-101],
...             [6,5,6,5,0,0,0,0],
...             [0,0,0,0,6,5,6,5],
...             [1,0,0,0,1,0,0,0],
...             [0,1,0,0,0,1,0,0],
...             [0,0,1,0,0,0,1,0],
...             [0,0,0,1,0,0,0,1]
...             ]
... rhs_ineq = [-609070,-624340,33222,39432,3980,6728,4101,2369]
... lhs_eq = [[1,1,1,1,0,0,0,0],[0,0,0,0,1,1,1,1]]
```

```
... rhs_eq = [5537,6572]
... opt_soltn = linprog(c=obj_func, A_ub=lhs_ineq,\
...                     b_ub=rhs_ineq, A_eq = lhs_eq,\
...                     b_eq = rhs_eq, method="highs")
... print(opt_soltn)
...
        message: Optimization terminated successfully. (HiGHS Status 7: Optimal)
        success: True
         status: 0
            fun: -750865.0199999999
              x: [ 0.000e+00  1.436e+03  4.101e+03  0.000e+00  0.000e+00
                   4.203e+03  0.000e+00  2.369e+03]
            nit: 19
          lower:  residual: [ 0.000e+00  1.436e+03  4.101e+03  0.000e+00
                              0.000e+00  4.203e+03  0.000e+00  2.369e+03]
                 marginals: [-2.842e-14  0.000e+00  0.000e+00 -2.842e-14
                             -2.842e-14  0.000e+00 -2.842e-14  0.000e+00]
          upper:  residual: [       inf        inf        inf        inf
                                   inf        inf        inf        inf]
                 marginals: [ 0.000e+00  0.000e+00  0.000e+00  0.000e+00
                              0.000e+00  0.000e+00  0.000e+00  0.000e+00]
          eqlin:  residual: [ 0.000e+00  0.000e+00]
                 marginals: [-6.450e+01 -5.991e+01]
        ineqlin:  residual: [ 2.973e+04  5.624e+04  1.436e+03  6.572e+03
                              3.980e+03  1.089e+03  0.000e+00  0.000e+00]
                 marginals: [-0.000e+00 -0.000e+00 -0.000e+00 -0.000e+00
                             -0.000e+00 -0.000e+00 -0.000e+00 -0.000e+00]
 mip_node_count: 0
 mip_dual_bound: 0.0
        mip_gap: 0.0
```

Python Pyomo scripts for Example 2.3

```
>>> # We are directly initializing the decision
... # variables, inputing the objective function
... # and constraints. Input efficiencies can
... # be achieved by writing input wrapper scripts.
... import pyomo.environ as pyo
...
... model = pyo.ConcreteModel()
... model.x_11 = pyo.Var(within=pyo.NonNegativeReals)
... model.x_21 = pyo.Var(within=pyo.NonNegativeReals)
... model.x_31 = pyo.Var(within=pyo.NonNegativeReals)
... model.x_41 = pyo.Var(within=pyo.NonNegativeReals)
... model.x_12 = pyo.Var(within=pyo.NonNegativeReals)
... model.x_22 = pyo.Var(within=pyo.NonNegativeReals)
... model.x_32 = pyo.Var(within=pyo.NonNegativeReals)
... model.x_42 = pyo.Var(within=pyo.NonNegativeReals)
...
... model.obj = pyo.Objective(\
...     expr= 64.5*model.x_11 + 64.5*model.x_21 + 64.5*model.x_31 + 64.5*model.\
x_41 \
...     + 59.91*model.x_12 + 59.91*model.x_22 + 59.91*model.x_32 + 59.91*model.\
x_42, sense=pyo.maximize)
...
... model.Con1 = pyo.Constraint(\
...     expr= 115*model.x_11 + 105*model.x_21 \
...     + 119*model.x_31 + 101*model.x_41 >= 609070)
```

```
... model.Con2 = pyo.Constraint(\
...     expr= 115*model.x_12 + 105*model.x_22 \
...     + 119*model.x_32 + 101*model.x_42 >= 624340)
... model.Con3 = pyo.Constraint(\
...     expr= 6*model.x_11 + 5*model.x_21 \
...     + 6*model.x_31 + 5*model.x_41 <= 33222)
... model.Con4 = pyo.Constraint(\
...     expr= 6*model.x_12 + 5*model.x_22 \
...     + 6*model.x_32 + 5*model.x_42 <= 39432)
... model.Con5 = pyo.Constraint(\
...     expr= model.x_11 + model.x_21 \
...     + model.x_31 + model.x_41 == 5537)
... model.Con6 = pyo.Constraint(\
...     expr= model.x_12 + model.x_22 \
...     + model.x_32 + model.x_42 == 6572)
... model.Con7 = pyo.Constraint(\
...     expr= model.x_11 + model.x_12 <= 3980)
... model.Con8 = pyo.Constraint(\
...     expr= model.x_21 + model.x_22 <= 6728)
... model.Con9 = pyo.Constraint(\
...     expr= model.x_31 + model.x_32 <= 4101)
... model.Con10 = pyo.Constraint(\
...     expr= model.x_41 + model.x_42 <= 2369)
...
... # create the solver interface and solve the model
... solver = pyo.SolverFactory('glpk')
... solver.solve(model)
... model.display()
...
Model unknown

  Variables:
    x_11 : Size=1, Index=None
        Key  : Lower : Value : Upper : Fixed : Stale : Domain
        None :     0 :   0.0 :  None : False : False : NonNegativeReals
    x_21 : Size=1, Index=None
        Key  : Lower : Value : Upper : Fixed : Stale : Domain
        None :     0 :   0.0 :  None : False : False : NonNegativeReals
    x_31 : Size=1, Index=None
        Key  : Lower : Value  : Upper : Fixed : Stale : Domain
        None :     0 : 3168.0 :  None : False : False : NonNegativeReals
    x_41 : Size=1, Index=None
        Key  : Lower : Value  : Upper : Fixed : Stale : Domain
        None :     0 : 2369.0 :  None : False : False : NonNegativeReals
    x_12 : Size=1, Index=None
        Key  : Lower : Value : Upper : Fixed : Stale : Domain
        None :     0 :   0.0 :  None : False : False : NonNegativeReals
    x_22 : Size=1, Index=None
        Key  : Lower : Value  : Upper : Fixed : Stale : Domain
        None :     0 : 6572.0 :  None : False : False : NonNegativeReals
    x_32 : Size=1, Index=None
        Key  : Lower : Value : Upper : Fixed : Stale : Domain
        None :     0 :   0.0 :  None : False : False : NonNegativeReals
    x_42 : Size=1, Index=None
        Key  : Lower : Value : Upper : Fixed : Stale : Domain
        None :     0 :   0.0 :  None : False : False : NonNegativeReals

  Objectives:
    obj : Size=1, Index=None, Active=True
        Key  : Active : Value
        None :   True : 750865.02

  Constraints:
    Con1 : Size=1
        Key  : Lower    : Body     : Upper
        None : 609070.0 : 616261.0 :  None
    Con2 : Size=1
```

```
        Key  : Lower     : Body      : Upper
        None : 624340.0 : 690060.0 :  None
    Con3 : Size=1
        Key  : Lower : Body     : Upper
        None :  None : 30853.0 : 33222.0
    Con4 : Size=1
        Key  : Lower : Body     : Upper
        None :  None : 32860.0 : 39432.0
    Con5 : Size=1
        Key  : Lower  : Body   : Upper
        None : 5537.0 : 5537.0 : 5537.0
    Con6 : Size=1
        Key  : Lower  : Body   : Upper
        None : 6572.0 : 6572.0 : 6572.0
    Con7 : Size=1
        Key  : Lower : Body : Upper
        None :  None :  0.0 : 3980.0
    Con8 : Size=1
        Key  : Lower : Body    : Upper
        None :  None : 6572.0 : 6728.0
    Con9 : Size=1
        Key  : Lower : Body    : Upper
        None :  None : 3168.0 : 4101.0
    Con10 : Size=1
        Key  : Lower : Body    : Upper
        None :  None : 2369.0 : 2369.0
```

Example 6.10. Here, we illustrate how to solve, using scipy.omtimize and Pyomo, respectively, the agricultural planning problem considered in Example 2.4.

Python scipy.optimize scripts for Example 2.4

```
>>> from scipy.optimize import linprog
... obj_func = [-14,-14,-14,-3,-3,-3,-20,-20,-20,-16,-16,-16]
... lhs_ineq = [[1,0,0,1,0,0,1,0,0,1,0,0],
...             [0,1,0,0,1,0,0,1,0,0,1,0],
...             [0,0,1,0,0,1,0,0,1,0,0,1],
...             [25,0,0,35,0,0,28,0,0,26,0,0],
...             [0,25,0,0,35,0,0,28,0,0,26,0],
...             [0,0,25,0,0,35,0,0,28,0,0,26],
...             [1,1,1,0,0,0,0,0,0,0,0,0],
...             [0,0,0,1,1,1,0,0,0,0,0,0],
...             [0,0,0,0,0,0,1,1,1,0,0,0],
...             [0,0,0,0,0,0,0,0,0,1,1,1]
...             ]
... rhs_ineq = [37,45,65,300,350,700,20,30,15,12]
... lhs_eq = [[1/37,-1/45,0,1/37,-1/45,0,1/37,-1/45,0,1/37,-1/45,0],
...           [0,-1/45,1/65,0,-1/45,1/65,0,-1/45,1/65,0,-1/45,1/65]]
... rhs_eq = [[0,0]]
... opt_soltn = linprog(c=obj_func, A_ub=lhs_ineq, \
...                     b_ub=rhs_ineq, A_eq = lhs_eq, \
...                     b_eq = rhs_eq, method="highs")
... print(opt_soltn)
...
        message: Optimization terminated successfully. (HiGHS Status 7: Optimal)
        success: True
         status: 0
            fun: -754.2666666666672
```

```
          x: [ 0.000e+00  1.400e+01  4.733e+00  0.000e+00  0.000e+00
               0.000e+00  0.000e+00  0.000e+00  1.500e+01  1.151e+01
               0.000e+00  4.889e-01]
        nit: 20
      lower:  residual: [ 0.000e+00  1.400e+01  4.733e+00  0.000e+00
                          0.000e+00  0.000e+00  0.000e+00  0.000e+00
                          1.500e+01  1.151e+01  0.000e+00  4.889e-01]
             marginals: [-1.776e-15  0.000e+00  0.000e+00  1.100e+01
                          2.929e+01  1.100e+01  7.105e-15  5.488e+00
                          0.000e+00  0.000e+00  1.829e+00  0.000e+00]
      upper:  residual: [       inf        inf        inf        inf
                                inf        inf        inf        inf
                                inf        inf        inf        inf]
             marginals: [ 0.000e+00  0.000e+00  0.000e+00  0.000e+00
                          0.000e+00  0.000e+00  0.000e+00  0.000e+00
                          0.000e+00  0.000e+00  0.000e+00  0.000e+00]
      eqlin:  residual: [ 0.000e+00  0.000e+00]
             marginals: [-5.180e+02 -9.100e+02]
    ineqlin:  residual: [ 2.549e+01  3.100e+01  4.478e+01  7.111e-01
                          0.000e+00  1.490e+02  1.267e+00  3.000e+01
                          0.000e+00  0.000e+00]
             marginals: [-0.000e+00 -0.000e+00 -0.000e+00 -0.000e+00
                         -1.829e+00 -0.000e+00 -0.000e+00 -0.000e+00
                         -6.000e+00 -2.000e+00]
mip_node_count: 0
mip_dual_bound: 0.0
       mip_gap: 0.0
```

Python Pyomo scripts for Example 2.4

```
>>> # We are directly initializing the decision
... # variables, inputing the objective function
... # and constraints. Input efficiencies can
... # be achieved by writing input wrapper scripts.
... import pyomo.environ as pyo
...
... model = pyo.ConcreteModel()
... model.x_11 = pyo.Var(within=pyo.NonNegativeReals)
... model.x_12 = pyo.Var(within=pyo.NonNegativeReals)
... model.x_13 = pyo.Var(within=pyo.NonNegativeReals)
... model.x_21 = pyo.Var(within=pyo.NonNegativeReals)
... model.x_22 = pyo.Var(within=pyo.NonNegativeReals)
... model.x_23 = pyo.Var(within=pyo.NonNegativeReals)
... model.x_31 = pyo.Var(within=pyo.NonNegativeReals)
... model.x_32 = pyo.Var(within=pyo.NonNegativeReals)
... model.x_33 = pyo.Var(within=pyo.NonNegativeReals)
... model.x_41 = pyo.Var(within=pyo.NonNegativeReals)
... model.x_42 = pyo.Var(within=pyo.NonNegativeReals)
... model.x_43 = pyo.Var(within=pyo.NonNegativeReals)
...
... myOb = 14*model.x_11 + 14*model.x_12 + 14*model.x_13 \
...      + 3*model.x_21 + 3*model.x_22 + 3*model.x_23 \
...      + 20*model.x_31 + 20*model.x_32 + 20*model.x_33 \
...      + 16*model.x_41 + 16*model.x_42 + 16*model.x_43
...
... model.obj = pyo.Objective(expr= myOb, sense=pyo.maximize)
...
... model.ULCon1 = pyo.Constraint(\
...      expr= model.x_11 + model.x_21 + model.x_31 + model.x_41 <= 37)
... model.ULCon2 = pyo.Constraint(\
```

```
...     expr= model.x_12 + model.x_22 + model.x_32 + model.x_42 <= 45)
... model.ULCon3 = pyo.Constraint(\
...     expr= model.x_13 + model.x_23 + model.x_33 + model.x_43 <= 65)
... model.WACon1 = pyo.Constraint(\
...     expr= 25*model.x_11 + 35*model.x_21 + 28*model.x_31 + 26*model.x_41 <= \
300)
... model.WACon2 = pyo.Constraint(\
...     expr= 25*model.x_12 + 35*model.x_22 + 28*model.x_32 + 26*model.x_42 <= \
350)
... model.WACon3 = pyo.Constraint(\
...     expr= 25*model.x_13 + 35*model.x_23 + 28*model.x_33 + 26*model.x_43 <= \
700)
... model.TACon1 = pyo.Constraint(\
...     expr= model.x_11 + model.x_12 + model.x_13 <= 20)
... model.TACon2 = pyo.Constraint(\
...     expr= model.x_21 + model.x_22 + model.x_23 <= 30)
... model.TACon3 = pyo.Constraint(\
...     expr= model.x_31 + model.x_32 + model.x_33 <= 15)
... model.TACon4 = pyo.Constraint(\
...     expr= model.x_41 + model.x_42 + model.x_43 <= 12)
...
... equalPropLHS1 = (1/37)*model.x_11+ (1/37)*model.x_21+ (1/37)*model.x_31 + (\
1/37)*model.x_41 \
...     - (1/45)*model.x_12 - (1/45)*model.x_22 - (1/45)*model.x_32 - (1/45)*mo\
del.x_42
...
... equalPropLHS2 = (1/65)*model.x_13+ (1/65)*model.x_23+ (1/65)*model.x_33 + (\
1/65)*model.x_43 \
... - (1/45)*model.x_12 - (1/45)*model.x_22 - (1/45)*model.x_32 - (1/45)*model.\
x_42
...
... model.EPCon1 = pyo.Constraint(expr = equalPropLHS1 == 0)
... model.EPCon2 = pyo.Constraint(expr = equalPropLHS2 == 0)
...
...
... # create the solver interface and solve the model
... solver = pyo.SolverFactory('glpk')
... solver.solve(model)
... model.display()
...
Model unknown

  Variables:
    x_11 : Size=1, Index=None
        Key  : Lower : Value : Upper : Fixed : Stale : Domain
        None :     0 :   0.0 :  None : False : False : NonNegativeReals
    x_12 : Size=1, Index=None
        Key  : Lower : Value : Upper : Fixed : Stale : Domain
        None :     0 :  14.0 :  None : False : False : NonNegativeReals
    x_13 : Size=1, Index=None
        Key  : Lower : Value            : Upper : Fixed : Stale : Domain
        None :     0 : 4.73333333333333 :  None : False : False : NonNegativeReals
    x_21 : Size=1, Index=None
        Key  : Lower : Value : Upper : Fixed : Stale : Domain
        None :     0 :   0.0 :  None : False : False : NonNegativeReals
    x_22 : Size=1, Index=None
        Key  : Lower : Value : Upper : Fixed : Stale : Domain
        None :     0 :   0.0 :  None : False : False : NonNegativeReals
    x_23 : Size=1, Index=None
        Key  : Lower : Value : Upper : Fixed : Stale : Domain
        None :     0 :   0.0 :  None : False : False : NonNegativeReals
    x_31 : Size=1, Index=None
        Key  : Lower : Value : Upper : Fixed : Stale : Domain
        None :     0 :   0.0 :  None : False : False : NonNegativeReals
    x_32 : Size=1, Index=None
        Key  : Lower : Value : Upper : Fixed : Stale : Domain
        None :     0 :   0.0 :  None : False : False : NonNegativeReals
```

```
  x_33 : Size=1, Index=None
      Key  : Lower : Value : Upper : Fixed : Stale : Domain
      None :     0 :  15.0 :  None : False : False : NonNegativeReals
  x_41 : Size=1, Index=None
      Key  : Lower : Value            : Upper : Fixed : Stale : Domain
      None :     0 : 11.5111111111111 :  None : False : False : NonNegativeReals
  x_42 : Size=1, Index=None
      Key  : Lower : Value : Upper : Fixed : Stale : Domain
      None :     0 :   0.0 :  None : False : False : NonNegativeReals
  x_43 : Size=1, Index=None
      Key  : Lower : Value             : Upper : Fixed : Stale : Domain
      None :     0 : 0.488888888888888 :  None : False : False : NonNegativeReals

Objectives:
  obj : Size=1, Index=None, Active=True
      Key  : Active : Value
      None :   True : 754.2666666666664

Constraints:
  ULCon1 : Size=1
      Key  : Lower : Body             : Upper
      None :  None : 11.5111111111111 :  37.0
  ULCon2 : Size=1
      Key  : Lower : Body : Upper
      None :  None : 14.0 :  45.0
  ULCon3 : Size=1
      Key  : Lower : Body               : Upper
      None :  None : 20.222222222222218 :  65.0
  WACon1 : Size=1
      Key  : Lower : Body              : Upper
      None :  None : 299.2888888888886 : 300.0
  WACon2 : Size=1
      Key  : Lower : Body  : Upper
      None :  None : 350.0 : 350.0
  WACon3 : Size=1
      Key  : Lower : Body              : Upper
      None :  None : 551.0444444444444 : 700.0
  TACon1 : Size=1
      Key  : Lower : Body              : Upper
      None :  None : 18.7333333333333 :  20.0
  TACon2 : Size=1
      Key  : Lower : Body : Upper
      None :  None :  0.0 :  30.0
  TACon3 : Size=1
      Key  : Lower : Body : Upper
      None :  None : 15.0 :  15.0
  TACon4 : Size=1
      Key  : Lower : Body               : Upper
      None :  None : 11.999999999999988 :  12.0
  EPCon1 : Size=1
      Key  : Lower : Body                   : Upper
      None :   0.0 : -2.7755575615628914e-16 :   0.0
  EPCon2 : Size=1
      Key  : Lower : Body                  : Upper
      None :   0.0 : -5.377642775528102e-17 :   0.0
```

Example 6.11. Here, we illustrate how to solve, using scipy.omtimize and Pyomo, respectively, the employee scheduling problem considered in Example 2.5.

Python scipy.optimize scripts for Example 2.5

```
>>> from scipy.optimize import linprog
... obj_func = [190,205,215,225,235]
... lhs_ineq = [[-1,0,0,0,0],
...             [-1,-1,0,0,0],
...             [-1,-1,0,0,0],
...             [-1,-1,-1,0,0],
...             [0,-1,-1,0,0],
...             [0,0,-1,-1,0],
...             [0,0,-1,-1,0],
...             [0,0,0,-1,0],
...             [0,0,0,-1,-1],
...             [0,0,0,0,-1]
...             ]
... rhs_ineq = [-48,-79,-65,-87,-64,-53,-62,-72,-25,-15]
... opt_soltn = linprog(c=obj_func, A_ub=lhs_ineq, b_ub=rhs_ineq, method="highs\
")
... print(opt_soltn)
...
        message: Optimization terminated successfully. (HiGHS Status 7: Optimal)
        success: True
         status: 0
            fun: 41965.0
              x: [ 4.800e+01  6.400e+01  0.000e+00  7.200e+01  1.500e+01]
            nit: 1
          lower:  residual: [ 4.800e+01  6.400e+01  0.000e+00  7.200e+01
                              1.500e+01]
                 marginals: [ 0.000e+00  0.000e+00  1.000e+01  0.000e+00
                              0.000e+00]
          upper:  residual: [       inf        inf        inf        inf
                                    inf]
                 marginals: [ 0.000e+00  0.000e+00  0.000e+00  0.000e+00
                              0.000e+00]
          eqlin:  residual: []
                 marginals: []
        ineqlin:  residual: [ 0.000e+00  3.300e+01  4.700e+01  2.500e+01
                              0.000e+00  1.900e+01  1.000e+01  0.000e+00
                              6.200e+01  0.000e+00]
                 marginals: [-1.900e+02 -0.000e+00 -0.000e+00 -0.000e+00
                             -2.050e+02 -0.000e+00 -0.000e+00 -2.250e+02
                             -0.000e+00 -2.350e+02]
 mip_node_count: 0
 mip_dual_bound: 0.0
        mip_gap: 0.0
```

Python Pyomo scripts for Example 2.5

```
>>> # We are directly initializing the decision
... # variables, inputing the objective function
... # and constraints.  Input efficiencies can
... # be achieved by writing input wrapper scripts.
... import pyomo.environ as pyo
...
```

```
... model = pyo.ConcreteModel()
... model.x_1 = pyo.Var(within=pyo.NonNegativeReals)
... model.x_2 = pyo.Var(within=pyo.NonNegativeReals)
... model.x_3 = pyo.Var(within=pyo.NonNegativeReals)
... model.x_4 = pyo.Var(within=pyo.NonNegativeReals)
... model.x_5 = pyo.Var(within=pyo.NonNegativeReals)
...
... model.obj = pyo.Objective(expr= 190*model.x_1+205*model.x_2  \
...                          + 215*model.x_3 + 225*model.x_4 \
...                          + 235*model.x_5 , sense=pyo.minimize)
...
... model.Con1 = pyo.Constraint(\
...     expr= model.x_1 >= 48)
... model.Con2 = pyo.Constraint(\
...     expr= model.x_1 + model.x_2 >= 79)
... model.Con3 = pyo.Constraint(\
...     expr= model.x_1 + model.x_2 >= 65)
... model.Con4 = pyo.Constraint(\
...     expr= model.x_1 + model.x_2 + model.x_3 >= 87)
... model.Con5 = pyo.Constraint(\
...     expr= model.x_2 + model.x_3 >= 64)
... model.Con6 = pyo.Constraint(\
...     expr= model.x_3 + model.x_4 >= 53)
... model.Con7 = pyo.Constraint(\
...     expr= model.x_3 + model.x_4 >= 62)
... model.Con8 = pyo.Constraint(\
...     expr= model.x_4 >= 72)
... model.Con9 = pyo.Constraint(\
...     expr= model.x_4 + model.x_5 >= 25)
... model.Con10 = pyo.Constraint(\
...     expr= model.x_5 >= 15)
...
... # create the solver interface and solve the model
... solver = pyo.SolverFactory('glpk')
... solver.solve(model)
... model.display()
...
Model unknown

  Variables:
    x_1 : Size=1, Index=None
        Key  : Lower : Value : Upper : Fixed : Stale : Domain
        None :     0 :  48.0 :  None : False : False : NonNegativeReals
    x_2 : Size=1, Index=None
        Key  : Lower : Value : Upper : Fixed : Stale : Domain
        None :     0 :  64.0 :  None : False : False : NonNegativeReals
    x_3 : Size=1, Index=None
        Key  : Lower : Value : Upper : Fixed : Stale : Domain
        None :     0 :   0.0 :  None : False : False : NonNegativeReals
    x_4 : Size=1, Index=None
        Key  : Lower : Value : Upper : Fixed : Stale : Domain
        None :     0 :  72.0 :  None : False : False : NonNegativeReals
    x_5 : Size=1, Index=None
        Key  : Lower : Value : Upper : Fixed : Stale : Domain
        None :     0 :  15.0 :  None : False : False : NonNegativeReals

  Objectives:
    obj : Size=1, Index=None, Active=True
        Key  : Active : Value
        None :   True : 41965.0

  Constraints:
    Con1 : Size=1
        Key  : Lower : Body : Upper
        None :  48.0 : 48.0 :  None
    Con2 : Size=1
        Key  : Lower : Body  : Upper
```

```
        None :   79.0 : 112.0 :   None
    Con3 : Size=1
        Key  : Lower : Body  : Upper
        None :   65.0 : 112.0 :   None
    Con4 : Size=1
        Key  : Lower : Body  : Upper
        None :   87.0 : 112.0 :   None
    Con5 : Size=1
        Key  : Lower : Body : Upper
        None :   64.0 : 64.0 :  None
    Con6 : Size=1
        Key  : Lower : Body : Upper
        None :   53.0 : 72.0 :  None
    Con7 : Size=1
        Key  : Lower : Body : Upper
        None :   62.0 : 72.0 :  None
    Con8 : Size=1
        Key  : Lower : Body : Upper
        None :   72.0 : 72.0 :  None
    Con9 : Size=1
        Key  : Lower : Body : Upper
        None :   25.0 : 87.0 :  None
    Con10 : Size=1
        Key  : Lower : Body : Upper
        None :   15.0 : 15.0 :  None
```

Exercises

6.1 Use scipy.optimize and/or a direct implementation of the Simplex Method to solve the minimization linear programming problem

- Minimize the linear objective function

$$Z = x_1 + 2x_2 + 3x_3$$

- Subject to the constraints

$$\begin{aligned} x_1 - 2x_2 + x_3 &\leq 1 \\ 3x_1 - 4x_2 + 6x_3 &\leq 8 \\ 2x_1 - 4x_2 + 2x_3 &\leq 2 \end{aligned}$$

and

$$x_1, x_2, x_3 \geq 0.$$

6.2 Use scipy.optimize and/or a direct implementation of the Simplex Method to solve the maximization linear programming problem

- Maximize the linear objective function

$$Z = -x_1 - x_2 - x_3$$

- Subject to the constraints

$$\begin{aligned} x_1 + 3x_2 \qquad &\leq 10 \\ x_1 + x_2 + x_3 &\geq 6 \\ x_1 \qquad\qquad &\geq 2 \end{aligned}$$

and

$$x_1, x_2, x_3 \geq 0.$$

6.3 Use scipy.optimize and/or a direct implementation of the Simplex Method to solve the minimization linear programming problem

- Minimize the linear objective function

$$Z = x_1 + 2x_2 + 3x_3$$

- Subject to the constraints

$$\begin{aligned} x_1 - 2x_2 + x_3 &\leq 1 \\ 3x_1 - 4x_2 + 6x_3 &\leq 8 \\ 2x_1 - 4x_2 + 2x_3 &\leq 2 \end{aligned}$$

and

$$x_1, x_2, x_3 \geq 0.$$

6.4 Use scipy.optimize and/or a direct implementation of the Simplex Method to solve the noncanonical maximization linear programming problem

- Maximize the linear objective function

$$Z = 2x_1 + 3x_2 + 4x_3$$

- Subject to the constraints

$$\begin{aligned} x_1 + x_2 + x_3 &\leq 1 \\ 3x_1 + 2x_2 + \tfrac{1}{2}x_3 &= 2 \\ 3x_1 + 2x_2 + x_3 &\geq 4 \end{aligned}$$

and

$$x_1, x_2, x_3 \geq 0.$$

6.5 Give a direct Python implementation of minimization Tucker Tableau and write a suitable Python script to convert such a minimization Tucker Tableau via the operation of negative transpose to a maximization Tucker Tableau.

6.6 Give a direct Python implementation of dual Tucker Tableau and write suitable Python scripts to allow for a direct implementation of the Simplex Method applied to such dual Tucker Tableau.

6.7 Develop efficient Python scripts and/or data structures to facilitate working with diet problems, mixture blending, agricultural planning and/or employee scheduling in scipy.optimize and/or Pyomo.

Chapter 7
Transportation and Assignment Problems

Overview

- Introduction to transportation problems.
- Introduction to assignment problems.

7.1 Transportation Problem: Motivational Example

In Example 7.1 below, we describe a representative instance of the *balanced transportation problem*. A detailed formulation of such transportation problems is given in Definition 7.2.

Example 7.1. Suppose that a manufacturer owns three warehouses and sells to three markets. A pictorial realization of the resulting transportation network is described in Figure 7.1.

The supply of each warehouse, the demand of each market and the shipping cost per ton of product from each warehouse to each market are summarized in Table 7.1.

In this context, the question that arises is how to ship the products so as to minimize the total transportation cost.

In order to formulate a solution to this problem, via a suitable linear mathematical programming problem, let x_{ij}, for $i, j = 1, 2, 3$, be the number of tons of product shipped from warehouse i to market j.

The aim is to

- Minimize the total transportation cost

$$C(x_{11}, \ldots, x_{33}) = 2x_{11} + x_{12} + 2x_{13} + 9x_{21} + 4x_{22} + 7x_{23} + x_{31} + 2x_{32} + 9x_{33}$$

N. Grieve, *An Introduction to Mathematical Programming and Network Science*,
Springer Undergraduate Texts in Mathematics and Technology,
https://doi.org/10.1007/978-3-032-13330-4_7

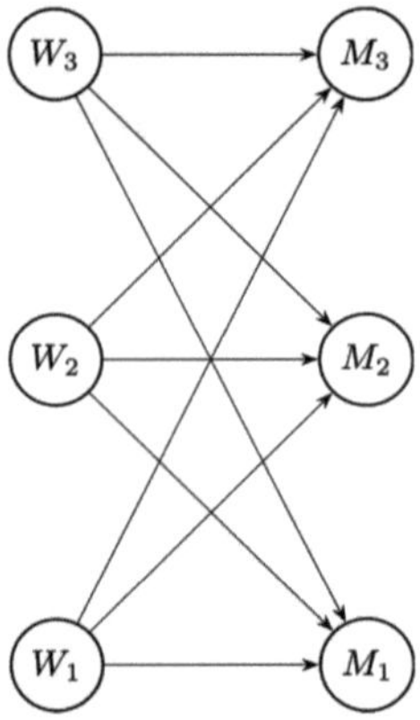

Fig. 7.1 Graphical representation of the transportation network for Example 7.1

Table 7.1 The warehouse market data table for Example 7.1

	Market 1	Market 2	Market 3	Warehouse supplies
Warehouse 1	\$2/ton	\$1/ton	\$2/ton	40 tons
Warehouse 2	\$9/ton	\$4/ton	\$7/ton	60 tons
Warehouse 3	\$1/ton	\$2/ton	\$9/ton	10 tons
Market demands	40 tons	50 tons	20 tons	110 tons

- Subject to
  - the warehouse constraints

$$\begin{aligned} x_{11} + x_{12} + x_{13} &= 40 \\ x_{21} + x_{22} + x_{23} &= 60 \\ x_{31} + x_{32} + x_{33} &= 10 \end{aligned}$$

  - the market demand constraints

$$\begin{aligned} x_{11} + x_{21} + x_{31} &= 40 \\ x_{12} + x_{22} + x_{32} &= 50 \\ x_{13} + x_{23} + x_{33} &= 20 \end{aligned}$$

  and
  - the nonnegativity constraints

$$x_{11}, x_{12}, x_{13}, x_{21}, x_{22}, x_{23}, x_{31}, x_{32}, x_{33} \geq 0.$$

It turns out that the corresponding minimization tableau is quite sparse. This suggests the fact that more specialized algorithms, in place of the Simplex Method, can be used to give solutions to such transportation problems. While this indeed turns out to be the case, here we restrict ourselves to using the Simplex Method directly to solve such transportation problems.

We also note that Algebraic Modelling Languages like Pyomo implement data structures which facilitate with studying and solving such transportation problems. We refer to Example 7.5 and/or the text [5] for more details.

Using any suitable implementation of the Simplex Method, for example, an optimal solution to the given transportation problem is found to be

$$C = 360.$$

This is achieved when $x_{11} = 30$, $x_{12} = 0$, $x_{13} = 10$, $x_{21} = 0$, $x_{22} = 50$, $x_{23} = 10$, $x_{31} = 10$, $x_{32} = 0$ and $x_{33} = 0$.

7.2 Formulation of Transportation Problems

As suggested in Section 7.1, one of the most basic *minimum cost network flow problems* is one in which each node is either a source (also called supply) or a sink (also called demand). For example, imagine a distributor with several warehouses and a group of customers. There is a cost that is associated with serving each customer from each given warehouse.

In this model, there are m supply nodes, each with an available supply s, and n demand nodes, each with a demand of d_j. We assume that the total supply in the network is equal to the total demand

$$\sum_{i=1}^{m} s_i = \sum_{j=1}^{n} d_j.$$

The objective is to satisfy all demands using the available supply and to accomplish this distribution using minimum cost routes. Such considerations give rise to the following formulation of the *transportation problem.*

Definition 7.2 (Transportation problem).

- Minimize the objective function

$$C = \sum_{i=1}^{m} \sum_{j=1}^{n} c_{ij} x_{ij}$$

- Subject to
  - the warehouse constraints

$$\sum_{j=1}^{n} x_{ij} = s_i \text{ for } i = 1, \ldots, m$$

  - the demand constraints

$$\sum_{i=1}^{m} x_{ij} = d_j \text{ for } j = 1, \ldots, n$$

  and
  - the nonnegativity constraints

$$x_{ij} \geq 0 \text{ for all } i \text{ and } j.$$

In formulating such transportation problems, a key underlying principle is that the total supply is equal to the total demand. That is

$$\begin{aligned} \text{total supply} &= \sum_{i=1}^{m} s_i = \sum_{i=1}^{m} \left(\sum_{j=1}^{n} x_{ij} \right) = \sum_{i=1}^{m} \sum_{j=1}^{n} x_{ij} \\ &= \sum_{i=1}^{m} \sum_{j=1}^{n} x_{ij} = \sum_{j=1}^{n} \left(\sum_{i=1}^{m} x_{ij} \right) \\ &= \sum_{j=1}^{n} d_j = \text{total demand.} \end{aligned}$$

The data needed to specify a transportation problem can be displayed in a *balanced transportation tableau*. See Table 7.2.

Table 7.2 Balanced transportation tableau

	M_1	...	M_n	
W_1	c_{11}	...	c_{1n}	s_1
$\vdots$	$\vdots$		$\vdots$	$\vdots$
W_m	c_{m1}	...	c_{mn}	s_m
	d_1	...	d_n	$\sum_{i=1}^{m} s_i = \sum_{j=1}^{n} d_j$

Such transportation problems can be solved via the Simplex Method and more specialized implementations thereof. In particular, such transportation problems of reasonable size can be solved via the linear programming features of scipy.optimize and/or other implementations of the Simplex Method and/or generalizations thereof. Further, Algebraic Modelling Languages like Pyomo implement abstract data types which facilitate working with and solving such balanced transportation problems.

7.3 Balanced Assignment Problem: Motivational Example

In Example 7.3 below, we describe a representative instance of the *balanced assignment problem*. We refer to Definition 7.4 for a precise formulation of such transportation problems.

Example 7.3. Consider the case that a company has purchased three new machines of different types. There are four available locations in the shop where a machine could be installed. Some of these locations are more desirable than others.

For example, they may have a close proximity to some work centres that will have heavy work flow from these machines. On the other hand, the handling of materials required to achieve a heavy work flow is very expensive. The objective is thus to assign the new machines to the available locations in such a way that minimizes the total cost of materials handling.

Suppose that the *estimated cost* in dollars per hour of materials handling is summarized by the entries in Table 7.3.

In the above table, Location 2 is deemed not suitable for Machine 2. Thus, there is no cost given for that case.

In order to model the situation as an *assignment problem*, there are a few steps. They are summarized in the following way.

- Introduce a "dummy machine" for the extra location. (So, we will have four machines and four locations.)
- Attach an extremely large cost M to the assignment of Machine 2 to Location 2 to prevent this assignment in the optimal solution.

Such considerations give rise to the following revised cost table, see Table 7.4, with dummy variables added for the given assignment problem.

Table 7.3 Assignment table for Example 7.3

	Location 1	Location 2	Location 3	Location 4
Machine 1	16	16	12	11
Machine 2	15	---	18	20
Machine 3	5	9	10	9

Table 7.4 Revised cost table for Example 7.3

	Task 1	Task 2	Task 3	Task 4
Assignee 1	16	16	12	11
Assignee 2	15	---	18	20
Assignee 3	5	9	10	9
Assignee 4 (D)	0	0	0	0

In Table 7.4, the idea is to view each location as a task and each machine as an assignee for the given assignment problem. Further, Table 7.4 contains all of the necessary data for solving the given assignment problem.

Given the above parameter table, a linear mathematical program that describes the given assignment problem can be formulated as

- Minimize the linear cost function

$$C = 16x_{11} + 16x_{12} + 12x_{13} + 11x_{14} \\ + 15x_{21} + Mx_{22} + 18x_{23} + 20x_{24} \\ 5x_{31} + 9x_{32} + 10x_{33} + 9x_{34}.$$

- Subject to the constraints

$$\sum_{j=1}^{4} x_{ij} = 1 \text{ for } i = 1, \ldots, 4$$

$$\sum_{i=1}^{4} x_{ij} = 1 \text{ for } j = 1, \ldots, 4$$

$$x_{ij} \geq 0 \text{ for } i, j = 1, \ldots, 4$$

and

$$x_{ij} \text{ are integers for } i, j = 1, \ldots, 4.$$

Here, we understand M to be some excessively high cost. In practice to solve the underlying linear programming problem, we assign M to be some sufficiently large positive number. Moreover, implicit in the above problem formulation is the fact that the decision variables will assume values of either 0 or 1. This is a key feature of assignment problems and reflects their inherent binary nature.

As it turns out, the optimal solution is to assign Machine 1 to Location 4, Machine 2 to Location 3 and Machine 3 to Location 1 for a total cost of $34/hour. In fact the "dummy machine" is assigned to Location 2; this location is available for some future real machine.

Indeed, these conclusions arise from the fact that the optimal solution to the given assignment problem is

$$C = 34$$

which is achieved when $x_{14} = 1$, $x_{23} = 1$, $x_{31} = 1$, $x_{42} = 1$ and all other decision variables $x_{ij} = 0$.

This optimal solution can be determined easily using the Simplex Method. Similarly to the case of transportation problems, Algebraic Modelling Languages like Pyomo implement data structures which facilitate with studying and solving such assignment problems. See Example 7.6 and/or the text [5, Chapter 3] for more details.

7.4 Formulation of Assignment Problems

Here, we give a general formulation of the *balanced assignment problem.* It can be seen as a special type of balanced transportation problem. Indeed, a general formulation of the *balanced assignment problem* is achieved in the following way.

Definition 7.4 (Balanced assignment problem).

- Minimize the cost function

$$C = \sum_{i=1}^{n} \sum_{j=1}^{n} c_{ij} x_{ij}$$

- Subject to the constraints

$$\sum_{j=1}^{n} x_{ij} = 1 \text{ for } i = 1, \ldots, n$$

$$\sum_{i=1}^{n} x_{ij} = 1 \text{ for } j = 1, \ldots, n$$

$$x_{ij} \geq 0 \text{ for all } i, j = 1, \ldots, n$$

and

$$x_{ij} \text{ are integers (either 0 or 1) for } i, j = 1, \ldots, n.$$

For our purposes, some key features to such a balanced assignment problem include the following considerations.

- There are n assignees that are required to be assigned to n tasks.
- Each of the n assignees is to be assigned to exactly one task.
- Each of the n tasks is to be performed by exactly one of the n assignees.

Table 7.5 Balanced assignment tableau

	J_1	…	J_n	
P_1	c_{11}	…	c_{1n}	1
⋮	⋮	⋮	⋮	⋮
P_n	c_{n1}	…	c_{nn}	1
	1	…	1	

- There is a cost c_{ij} that is associated with assignee i, for $i = 1, \dots, n$ for performing task j, for $j = 1, \dots, n$.
- The objective is to determine how all of the n assignments should be made to minimize the total cost.
- In each optimal solution, the decision variables x_{ij} will take on binary values. Especially, there is no need to explicitly state this as an additional constraint. A proof of this inherent feature of balanced assignment problems can be found, for instance, in [13, Section 6.5]. As a consequence of this phenomenon such balanced assignment problems can be solved directly via the Simplex Method.

As in the case of transportation problems, the data of a balanced assignment problem can be represented using a *balanced assignment tableau* as described in Table 7.5.

Such balanced assignment problems fall within the realm of what are sometimes called *binary linear programming problems*. They can be solved via the Simplex Method and more specialized implementations thereof. The adjective *binary* arises from the fact that the decision variables are required to be *binary integers*—that is they are required to be either 0 or 1.

For practical purposes, we note that such balanced assignment problems of reasonable size can be solved easily by using the Simplex Method.

7.5 Selected Examples

In Examples 7.5 and 7.6 below, we indicate how to solve Examples 7.1 and 7.3 using the Algebraic Modelling Language Pyomo. The scripts that we provide here in Examples 7.1 and 7.3 can be adapted to solve Examples 7.7–7.9.

All of these examples which we describe here, can be solved using the scipy.optimize package. Moreover, the approach that we give using Pyomo is direct and without making use of any auxiliary constructs. To assist with making the input of such problems more efficient, techniques such as those illustrated in [5, Section 3.3] can be employed.

Example 7.5. Included below are a collection of Pyomo scripts which illustrate how to use Pyomo to solve Example 7.1.

Python Pyomo scripts for Example 7.1

```
>>> # We are directly initializing the decision
... # variables, inputing the objective function
... # and constraints. Input efficiencies can
... # be achieved by writing input wrapper scripts.
... import pyomo.environ as pyo
...
... model = pyo.ConcreteModel()
... model.x_11 = pyo.Var(within=pyo.NonNegativeReals)
... model.x_12 = pyo.Var(within=pyo.NonNegativeReals)
... model.x_13 = pyo.Var(within=pyo.NonNegativeReals)
... model.x_21 = pyo.Var(within=pyo.NonNegativeReals)
... model.x_22 = pyo.Var(within=pyo.NonNegativeReals)
... model.x_23 = pyo.Var(within=pyo.NonNegativeReals)
... model.x_31 = pyo.Var(within=pyo.NonNegativeReals)
... model.x_32 = pyo.Var(within=pyo.NonNegativeReals)
... model.x_32 = pyo.Var(within=pyo.NonNegativeReals)
... model.x_33 = pyo.Var(within=pyo.NonNegativeReals)
... model.obj = pyo.Objective(\
...     expr= 2*model.x_11 + model.x_12 + 2*model.x_13 \
...     + 9*model.x_21 + 4*model.x_22 + 7*model.x_23 \
...     + model.x_31+2*model.x_32+9*model.x_33,\
...     sense=pyo.minimize)
... model.WarehouseCon1 = pyo.Constraint(\
...     expr= model.x_11 + model.x_12 + model.x_13 == 40)
... model.WarehouseCon2 = pyo.Constraint(\
...     expr= model.x_21 + model.x_22 + model.x_23 == 60)
... model.WarehouseCon3 = pyo.Constraint(\
...     expr= model.x_31 + model.x_32 + model.x_33 == 10)
... model.DistributionCon1 = pyo.Constraint(\
...     expr= model.x_11 + model.x_21 + model.x_31 == 40)
... model.DistributionCon2 = pyo.Constraint(\
...     expr= model.x_12 + model.x_22 + model.x_32 == 50)
... model.DistributionCon3 = pyo.Constraint(\
...     expr= model.x_13 + model.x_23 + model.x_33 == 20)
...
... # create the solver interface and solve the model
... solver = pyo.SolverFactory('glpk')
... solver.solve(model)
... model.display()
...
Model unknown

  Variables:
    x_11 : Size=1, Index=None
        Key  : Lower : Value : Upper : Fixed : Stale : Domain
        None :     0 :  30.0 :  None : False : False : NonNegativeReals
    x_12 : Size=1, Index=None
        Key  : Lower : Value : Upper : Fixed : Stale : Domain
        None :     0 :   0.0 :  None : False : False : NonNegativeReals
    x_13 : Size=1, Index=None
        Key  : Lower : Value : Upper : Fixed : Stale : Domain
        None :     0 :  10.0 :  None : False : False : NonNegativeReals
    x_21 : Size=1, Index=None
        Key  : Lower : Value : Upper : Fixed : Stale : Domain
        None :     0 :   0.0 :  None : False : False : NonNegativeReals
    x_22 : Size=1, Index=None
        Key  : Lower : Value : Upper : Fixed : Stale : Domain
        None :     0 :  50.0 :  None : False : False : NonNegativeReals
    x_23 : Size=1, Index=None
        Key  : Lower : Value : Upper : Fixed : Stale : Domain
        None :     0 :  10.0 :  None : False : False : NonNegativeReals
    x_31 : Size=1, Index=None
        Key  : Lower : Value : Upper : Fixed : Stale : Domain
```

```
        None :      0 :  10.0 :  None : False : False : NonNegativeReals
    x_32 : Size=1, Index=None
        Key  : Lower : Value : Upper : Fixed : Stale : Domain
        None :      0 :   0.0 :  None : False : False : NonNegativeReals
    x_33 : Size=1, Index=None
        Key  : Lower : Value : Upper : Fixed : Stale : Domain
        None :      0 :   0.0 :  None : False : False : NonNegativeReals

  Objectives:
    obj : Size=1, Index=None, Active=True
        Key  : Active : Value
        None :   True : 360.0

  Constraints:
    WarehouseCon1 : Size=1
        Key  : Lower : Body : Upper
        None :  40.0 : 40.0 :  40.0
    WarehouseCon2 : Size=1
        Key  : Lower : Body : Upper
        None :  60.0 : 60.0 :  60.0
    WarehouseCon3 : Size=1
        Key  : Lower : Body : Upper
        None :  10.0 : 10.0 :  10.0
    DistributionCon1 : Size=1
        Key  : Lower : Body : Upper
        None :  40.0 : 40.0 :  40.0
    DistributionCon2 : Size=1
        Key  : Lower : Body : Upper
        None :  50.0 : 50.0 :  50.0
    DistributionCon3 : Size=1
        Key  : Lower : Body : Upper
        None :  20.0 : 20.0 :  20.0
```

Example 7.6. Included below are a collection of Pyomo scripts which illustrate how to use Pyomo to solve Example 7.3.

Python Pyomo scripts for Example 7.3

```
>>> # We are directly initializing the decision
... # variables, inputing the objective function
... # and constraints. Input efficiencies can
... # be achieved by writing input wrapper scripts.
... import pyomo.environ as pyo
...
... model = pyo.ConcreteModel()
... model.x_11 = pyo.Var(within=pyo.Binary)
... model.x_12 = pyo.Var(within=pyo.Binary)
... model.x_13 = pyo.Var(within=pyo.Binary)
... model.x_14 = pyo.Var(within=pyo.Binary)
... model.x_21 = pyo.Var(within=pyo.Binary)
... model.x_22 = pyo.Var(within=pyo.Binary)
... model.x_23 = pyo.Var(within=pyo.Binary)
... model.x_24 = pyo.Var(within=pyo.Binary)
... model.x_31 = pyo.Var(within=pyo.Binary)
... model.x_32 = pyo.Var(within=pyo.Binary)
... model.x_33 = pyo.Var(within=pyo.Binary)
... model.x_34 = pyo.Var(within=pyo.Binary)
... model.x_41 = pyo.Var(within=pyo.Binary)
... model.x_42 = pyo.Var(within=pyo.Binary)
```

```
... model.x_43 = pyo.Var(within=pyo.Binary)
... model.x_44 = pyo.Var(within=pyo.Binary)
...
... model.obj = pyo.Objective(\
...     expr= 16*model.x_11 + 16*model.x_12 + 12*model.x_13 + 11*model.x_14 \
...     + 15*model.x_21 + 100*model.x_22+18*model.x_23+20*model.x_24 \
...     +5*model.x_31+9*model.x_32+10*model.x_33 + 9*model.x_34, \
...     sense=pyo.minimize)
... model.RowCon1 = pyo.Constraint(\
...     expr= model.x_11 + model.x_12 + model.x_13 + model.x_14 == 1)
... model.RowCon2 = pyo.Constraint(\
...     expr= model.x_21 + model.x_22 + model.x_23 + model.x_24 == 1)
... model.RowCon3 = pyo.Constraint(\
...     expr= model.x_31 + model.x_32 + model.x_33 + model.x_34 == 1)
... model.RowCon4 = pyo.Constraint(\
...     expr= model.x_41 + model.x_42 + model.x_43 + model.x_44 == 1)
... model.ColCon5 = pyo.Constraint(\
...     expr= model.x_11 + model.x_21 + model.x_31 + model.x_41 == 1)
... model.ColCon6 = pyo.Constraint(\
...     expr= model.x_12 + model.x_22 + model.x_32 + model.x_42 == 1)
... model.ColCon7 = pyo.Constraint(\
...     expr= model.x_13 + model.x_23 + model.x_33 + model.x_43 == 1)
... model.ColCon8 = pyo.Constraint(\
...     expr= model.x_14 + model.x_24 + model.x_34 + model.x_44 == 1)
...
... # create the solver interface and solve the model
... solver = pyo.SolverFactory('glpk')
... solver.solve(model)
... model.display()
...
Model unknown

  Variables:
    x_11 : Size=1, Index=None
        Key  : Lower : Value : Upper : Fixed : Stale : Domain
        None :     0 :   0.0 :     1 : False : False : Binary
    x_12 : Size=1, Index=None
        Key  : Lower : Value : Upper : Fixed : Stale : Domain
        None :     0 :   0.0 :     1 : False : False : Binary
    x_13 : Size=1, Index=None
        Key  : Lower : Value : Upper : Fixed : Stale : Domain
        None :     0 :   0.0 :     1 : False : False : Binary
    x_14 : Size=1, Index=None
        Key  : Lower : Value : Upper : Fixed : Stale : Domain
        None :     0 :   1.0 :     1 : False : False : Binary
    x_21 : Size=1, Index=None
        Key  : Lower : Value : Upper : Fixed : Stale : Domain
        None :     0 :   0.0 :     1 : False : False : Binary
    x_22 : Size=1, Index=None
        Key  : Lower : Value : Upper : Fixed : Stale : Domain
        None :     0 :   0.0 :     1 : False : False : Binary
    x_23 : Size=1, Index=None
        Key  : Lower : Value : Upper : Fixed : Stale : Domain
        None :     0 :   1.0 :     1 : False : False : Binary
    x_24 : Size=1, Index=None
        Key  : Lower : Value : Upper : Fixed : Stale : Domain
        None :     0 :   0.0 :     1 : False : False : Binary
    x_31 : Size=1, Index=None
        Key  : Lower : Value : Upper : Fixed : Stale : Domain
        None :     0 :   1.0 :     1 : False : False : Binary
    x_32 : Size=1, Index=None
        Key  : Lower : Value : Upper : Fixed : Stale : Domain
        None :     0 :   0.0 :     1 : False : False : Binary
    x_33 : Size=1, Index=None
        Key  : Lower : Value : Upper : Fixed : Stale : Domain
        None :     0 :   0.0 :     1 : False : False : Binary
    x_34 : Size=1, Index=None
```

```
      Key  : Lower : Value : Upper : Fixed : Stale : Domain
      None :     0 :   0.0 :     1 : False : False : Binary
  x_41 : Size=1, Index=None
      Key  : Lower : Value : Upper : Fixed : Stale : Domain
      None :     0 :   0.0 :     1 : False : False : Binary
  x_42 : Size=1, Index=None
      Key  : Lower : Value : Upper : Fixed : Stale : Domain
      None :     0 :   1.0 :     1 : False : False : Binary
  x_43 : Size=1, Index=None
      Key  : Lower : Value : Upper : Fixed : Stale : Domain
      None :     0 :   0.0 :     1 : False : False : Binary
  x_44 : Size=1, Index=None
      Key  : Lower : Value : Upper : Fixed : Stale : Domain
      None :     0 :   0.0 :     1 : False : False : Binary

Objectives:
  obj : Size=1, Index=None, Active=True
      Key  : Active : Value
      None :   True :  34.0

Constraints:
  RowCon1 : Size=1
      Key  : Lower : Body : Upper
      None :   1.0 :  1.0 :   1.0
  RowCon2 : Size=1
      Key  : Lower : Body : Upper
      None :   1.0 :  1.0 :   1.0
  RowCon3 : Size=1
      Key  : Lower : Body : Upper
      None :   1.0 :  1.0 :   1.0
  RowCon4 : Size=1
      Key  : Lower : Body : Upper
      None :   1.0 :  1.0 :   1.0
  ColCon5 : Size=1
      Key  : Lower : Body : Upper
      None :   1.0 :  1.0 :   1.0
  ColCon6 : Size=1
      Key  : Lower : Body : Upper
      None :   1.0 :  1.0 :   1.0
  ColCon7 : Size=1
      Key  : Lower : Body : Upper
      None :   1.0 :  1.0 :   1.0
  ColCon8 : Size=1
      Key  : Lower : Body : Upper
      None :   1.0 :  1.0 :   1.0
```

Example 7.7. Here, we formulate the transportation problem that is associated to the transportation tableau given in Table 7.6 below.

Table 7.6 Transportation data table for Example 7.7

	M_1	M_2	M_3	M_4	
W_1	7	12	8	50	24
W_2	11	6	10	8	22
W_3	14	50	5	9	30
	15	18	26	17	76

Table 7.7 Transportation data tableau for Example 7.7

	M_1	M_2	M_3	M_4				M_1	M_2	M_3	M_4	
W_1	7	12	8	50	24		W_1	c_{11}	c_{12}	c_{13}	c_{14}	s_1
W_2	11	6	10	8	22	=	W_2	c_{21}	c_{22}	c_{23}	c_{24}	s_2
W_3	14	50	5	9	30		W_3	c_{23}	c_{23}	c_{33}	c_{34}	s_3
	15	18	26	17	76			d_1	d_2	d_3	d_4	$\sum_i s_i = \sum_j d_j$

We first fix our notational conventions as in Table 7.7.
The relevant transportation problem thus takes the form

- Minimize the linear objective function

$$C = 7x_{11} + 12x_{12} + 8x_{12} + 50x_{12} \\ + 11x_{21} + 6x_{22} + 10x_{23} + 8x_{24} \\ + 14x_{31} + 50x_{32} + 5x_{33} + 9x_{34}$$

- Subject to the supply constraints

$$\begin{aligned} x_{11} + x_{12} + x_{13} + x_{14} &= 24 \\ x_{21} + x_{22} + x_{23} + x_{24} &= 22 \\ x_{31} + x_{32} + x_{33} + x_{34} &= 30 \end{aligned}$$

- the demand constraints

$$\begin{aligned} x_{11} + x_{21} + x_{31} &= 15 \\ x_{12} + x_{22} + x_{32} &= 18 \\ x_{13} + x_{23} + x_{33} &= 26 \\ x_{14} + x_{24} + x_{34} &= 17 \end{aligned}$$

and
- subject to the nonnegativity constraints

$$x_{ij} \geq 0 \text{ for all } i, j.$$

Using the Simplex Method an optimal solution is then found to be

$$Z = 519$$

which is achieved when $x_{11} = 15$, $x_{31} = 9$, $x_{22} = 18$, $x_{24} = 4$, $x_{33} = 17$, $x_{34} = 13$ and all other x_{ij} equal to zero.

Table 7.8 Assignment data table for Example 7.8

	J_1	J_2	J_3	
P_1	8	7	15	1
P_2	5	8	11	1
P_3	4	5	7	1
	1	1	1	

Example 7.8. The assignment problem that is associated to the assignment tableau described in Table 7.8 is formulated as

- Minimize the cost function

$$C = 8x_{11} + 7x_{12} + 15x_{13} + 5x_{21} + 8x_{22} + 11x_{23} + 4x_{31} + 5x_{32} + 7x_{33}$$

- Subject to the constraints

$$\begin{aligned} x_{11} + x_{12} + x_{13} &= 1 \\ x_{21} + x_{22} + x_{23} &= 1 \\ x_{31} + x_{32} + x_{33} &= 1 \\ x_{11} + x_{21} + x_{31} &= 1 \\ x_{12} + x_{22} + x_{32} &= 1 \\ x_{13} + x_{23} + x_{33} &= 1 \end{aligned}$$

and

$$x_{ij} = 0 \text{ or } 1 \text{ for } i, j = 1, \ldots, n.$$

Using the Simplex Method an optimal solution is then found to be

$$C = 19$$

which is achieved when $x_{12} = 1$, $x_{21} = 1$, $x_{33} = 1$ and all other $x_{ij} = 0$.

Example 7.9. A plastics manufacturer has 1200 boxes of plastic wrap in stock at one factory and another 1000 boxes at its second factory. The manufacturer has orders for this product from three different retailers in quantities of 1000, 700 and 500 boxes respectively. The unit shipping costs (in cents per box) from the factories to the retailers are given in Table 7.9.

Table 7.9 Unit shipping cost data table for Example 7.9

	Retailer 1	Retailer 2	Retailer 3
Factory 1	17	13	11
Factory 2	13	9	12

The aim is to determine a minimum cost shipping schedule satisfying all demands from current inventory.

We can formulate a solution as an integer transportation problem. Indeed, let x_{ij} for $i = 1, 2$ and $j = 1, 2, 3$ be the number of boxes to be shipped from factory i to retailer j. We want to

- Minimize the objective function

$$C = 17x_{11} + 13x_{12} + 11x_{13} + 13x_{21} + 9x_{22} + 12x_{23}$$

- Subject to the warehouse constraints

$$\begin{aligned} x_{11} + x_{12} + x_{13} &= 1200 \\ x_{21} + x_{22} + x_{23} &= 1000 \end{aligned}$$

- the market constraints

$$\begin{aligned} x_{11} + x_{21} &= 1000 \\ x_{12} + x_{22} &= 700 \\ x_{13} + x_{23} &= 500 \end{aligned}$$

and

- the nonnegativity and integrality constraints

$$x_{ij} \geq 0 \text{ and integral for } i = 1, 2 \text{ and } j = 1, 2, 3.$$

The balanced transportation problem for this integer balanced transportation problem is given in Table 7.10.

Using the Simplex Method, an optimal solution is found to be

$$C = 27600$$

which is achieved when $x_{11} = 700$, $x_{31} = 500$, $x_{21} = 300$, $x_{22} = 700$ and all other $x_{ij} = 0$.

Table 7.10 Integer balanced transportation problem data table for Example 7.9

	M_1	M_2	M_3	
W_1	17	13	11	1200
W_2	13	9	12	1000
	1000	700	500	2200

Exercises

7.1 Formulate and solve the transportation problem that is associated to the transportation tableau that is given in Table 7.11.

7.2 Formulate and solve the transportation problem that is associated to the transportation tableau that is described in Table 7.12.

7.3 In the following transportation tableau, Table 7.13, x denotes a prohibitively high cost

Formulate and solve the corresponding transportation problem.

7.4 Suppose that a company wishes to assign six of its workers to six different jobs (with one worker assigned to each job and vice versa). The rating of each worker with respect to each job on a scale of 0–10 (with 10 being a high rating) is given in Table 7.14.

If the company wishes to maximize the total rating of the job assignment, then find the optimal assignment plan and the corresponding maximum total job assignment rating.

Table 7.11 Data table for Exercise 7.1

	M_1	M_2	M_3	
W_1	7	2	4	10
W_2	10	5	9	20
W_3	7	3	5	30
	20	10	30	

Table 7.12 Data table for Exercise 7.2

	M_1	M_2	M_3	M_4	M_5	
W_1	10	20	15	6	0	15
W_2	26	30	30	20	16	10
W_3	28	29	25	13	8	15
W_4	15	20	25	5	5	16
	9	15	9	15	8	

Table 7.13 Data table for Exercise 7.3

	M_1	M_2	M_3	
W_1	6	9	-1	20
W_2	3	9	11	15
W_3	9	1	x	20
W_4	x	6	7	15
	20	20	30	

Table 7.14 Data table for Exercise 7.4

	J_1	J_2	J_3	J_4	J_5	J_6
W_1	8	9	6	3	7	5
W_2	4	3	9	7	5	6
W_3	7	3	2	1	9	7
W_4	4	6	5	4	8	5
W_5	7	6	1	2	8	7
W_6	4	4	5	4	5	7

Table 7.15 Data table for Exercise 7.5

	1	2	3	4
A	8	6	5	7
B	6	5	3	4
C	7	8	4	6
D	6	7	5	6

7.5 Consider the assignment problem that has cost table given in Table 7.15.

In the table, the tasks are denoted by $1, 2, 3, 4$, whereas the assignees are denoted by A, B, C, D.

(i) Draw a network representation of this assignment problem.
(ii) Formulate this assignment problem as a suitable integer transportation problem and construct a suitable transportation tableau.
(iii) Determine an optimal solution.

7.6 Suppose that four cargo ships will be used for shipping goods from one port to four other ports labelled $1, 2, 3, 4$. Suppose that each ship can be used for making any one of these four trips. However, because of differences in the ships and cargos, suppose that the total cost of loading, transporting and unloading the goods for the different ship-port combinations varies considerably.

The relevant details are summarized in Table 7.16.

Table 7.16 Data table for Exercise 7.6

	1	2	3	4
A	500	400	600	700
B	600	600	700	500
C	700	500	700	600
D	500	400	600	600

In the above table, the ports are labelled 1, 2, 3, 4, whereas the ships are labelled A, B, C, D. The objective is to assign the four ships to four different ports in such a way as to minimize the total cost for all four shipments.

(i) Formulate an appropriate assignment problem and find an optimal solution.
(ii) Reformulate this problem as a suitable integer transportation problem.

Chapter 8
Network Flow Problems

Overview

- Introduction to network flow problems.
- Capacitated networks.
- Maximal flow network problem.
- Shortest path network problem.
- Minimal cost flow network problem.

8.1 Motivational Considerations for Network Flow Problems

Network flow problems are a certain kind of linear programming problems that arise in *graph theoretic contexts*. It turns out that the *transportation* and *assignment problems*, as formulated in Definitions 7.2 and 7.4, respectively, admit a reformulation as network flow problems.

In Sections 8.3, 8.5 and 8.6, we introduce, in precise terms, the network flow problems that we will consider. In Example 8.1 below, we discuss a motivational example. In Sections 8.2 and 8.4, we document relevant graph theoretic preliminaries.

Example 8.1. Consider the case of a directed road network in which each directed road is subject to *capacity limits*. For example, due to weight restrictions, perhaps each directed road segment can have only at most a certain fixed number of vehicles on it at any particular time.

During high season, the road network undergoes increased usage. Thus, various routes may be taken regardless of the distance so as to increase the number of trips that can be made in each given day. The network flow problem that arises within this context is the question of how to maximize the number of trips that can be made without violating the capacity limits on any given individual directed route.

N. Grieve, *An Introduction to Mathematical Programming and Network Science*,
Springer Undergraduate Texts in Mathematics and Technology,
https://doi.org/10.1007/978-3-032-13330-4_8

Table 8.1 Table of capacity constraints for Example 8.1

	v_0	v_1	v_2	v_3	v_4	v_5
v_0	0	6	0	10	0	0
v_1	0	0	3	0	5	0
v_2	0	0	0	0	4	9
v_3	0	3	2	0	5	0
v_4	0	0	0	0	0	9
v_5	0	0	0	0	0	0

In more concrete terms, pictured below is a graphical representation of a certain capacitated network together which has *source* $v_s = v_0$ and *destination* $v_d = v_5$.

The edge weights indicate the capacity constraints. They are given in Table 8.1.

In Table 8.1, the i, j entry c_{ij} is the *edge weight* from the edge v_i to the edge v_j. In particular, $c_{ij} = 0$ if there is no edge from vertex v_i to vertex v_j.

In the formulation of the underlying network flow problem, the idea is to introduce a decision variable x_{ij} for each i, j such that $c_{ij} > 0$. Then the underlying maximal network flow problem with capacity constraints can be modelled as the following maximization linear programming problem.

- Maximize the linear objective function

$$f = x_{25} + x_{45}$$

- Subject to the constraints

$$\begin{aligned} x_{01} + x_{31} - x_{12} - x_{14} &= 0 \\ x_{12} + x_{32} - x_{24} - x_{25} &= 0 \\ x_{03} - x_{31} - x_{34} - x_{32} &= 0 \\ x_{34} + x_{14} + x_{24} - x_{45} &= 0 \end{aligned}$$

$$x_{01} \leq 6, x_{03} \leq 10, x_{12} \leq 3, x_{14} \leq 5, x_{24} \leq 4$$

$$x_{25} \leq 9, x_{31} \leq 3, x_{32} \leq 2, x_{34} \leq 5, x_{45} \leq 9$$

and

$$x_{ij} \geq 0 \text{ for all } i, j.$$

Using the Simplex Method, an optimal solution to this network maximal flow linear programming problem is found to be

$$f = 14$$

which is achieved when $x_{01} = 6, x_{03} = 8, x_{12} = 3, x_{14} = 4, x_{24} = 0, x_{25} = 5, x_{31} = 1, x_{32} = 2, x_{34} = 5, x_{45} = 9$.

8.2 Directed Graphs, Capacity Flow Constraints and Conservation of Flow

In this section, we fix terminology which we use to discuss features of *directed networks* and *flows* on such networks with or without the presence of *capacity constraints*.

Definition 8.2.

- A *directed network* or *directed graph* $N = (V, E)$ consists of a finite nonempty set V whose elements are called vertices and a set E of ordered pairs of elements of V; the elements of E are called (directed) edges.
- A directed network N is *capacitated* if to each edge $(v_i, v_j) \in E$ there corresponds a real number $c_{ij} \geq 0$, which is called its *capacity*.
- A *flow* in a capacitated directed network N is an assignment of a real number x_{ij} to each edge $(v_i, v_j) \in E$ such that $0 \leq x_{ij} \leq c_{ij}$.
- The *decision variables* of a directed network are x_{ij} for $(v_i, v_j) \in E$. Thus, x_{ij} represents the flow numbers at the edge $(v_i, v_j) \in E$.

In Definition 8.3, we define the concept of *net input flow* at each given vertex. This turns out to be the sum of all flow numbers on edges into the vertex minus the sum of all flow numbers on edges out of the vertex. A key underlying principle is the idea of *conservation of network flow*. This is made precise in Theorem 8.4.

Definition 8.3. Let $N = (V, E)$ be a *capacitated directed network* with a given *flow*.

- If $v_j \in V$, then the *net input flow at vertex* v_j, denoted by $\phi(v_j)$, is

$$\phi(v_j) = \sum_i x_{ij} - \sum_i x_{ji}.$$

- If $\phi(v_j) > 0$, then v_j is said to be a *destination*.
- If $\phi(v_j) < 0$, then v_j is said to be a *source*.
- If $\phi(v_j) = 0$, then v_j is said to be an *intermediate vertex*.

To emphasize, in Definition 8.3, observe that the net input flow at a vertex is the sum of all flow numbers on edges into the vertex minus the sum of all flow numbers on edges out of the vertex. Another consideration is the *principle of conservation of network flow* which ensures that the total sum of the net input flow at each particular vertex is zero. This is made precise in Theorem 8.4 below.

Theorem 8.4 (Conservation of Flow). *Each flow in a capacitated directed network $N = (V, E)$ has the property that $\sum_j \phi(v_j) = 0$.*

Proof. By definition

$$\phi(v_j) = \sum_i x_{ij} - \sum_i x_{ji} \text{ for all } v_j \in V.$$

Therefore

$$\sum_j \phi(v_j) = \sum_j \left(\sum_i x_{ij} - \sum_i x_{ji} \right) = \sum_j \sum_i x_{ij} - \sum_j \sum_i x_{ji}. \tag{8.1}$$

The desired conclusion

$$\sum_j \phi(v_j) = 0$$

then follows since each of the network flow numbers $x_{\ell k}$, for $(v_\ell, v_k) \in E$ appears exactly once (and with different signs) in each of the two double summations that arise in the right-hand side of (8.1). □

Example 8.5. For the directed network that is given in Example 8.1, and described in Figure 8.1, we can consider the network flow capacity table which is given in Table 8.2.

In this table, each i, j entry (c_{ij}, x_{ij}) specifies a pair c_{ij} which represents the capacity of the edge from vertex v_i to vertex v_j, whereas x_{ij} specifies the given flow along this directed edge. Zero entries indicate that there is no edge from v_i to v_j.

The flow that is specified by the (nonzero) decision variables assuming the values $x_{01} = 6, x_{03} = 8, x_{12} = 3, x_{14} = 5, x_{25} = 5, x_{31} = 2, x_{32} = 2, x_{34} = 4, x_{45} = 9$ turns out to be an optimal flow as shown in Example 8.1. In particular, the objective function assumes the optimal maximal value for these values of the decision variables.

The corresponding net input flow at each of the vertices $v_0, \ldots, v_5$ are described in Table 8.3.

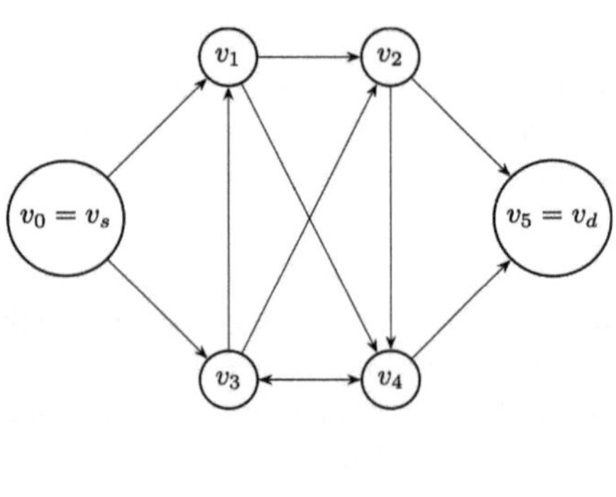

Fig. 8.1 Graphical representation of the directed network for Examples 8.5 and 8.7

Table 8.2 Network flow capacity data table for Example 8.5

	v_0	v_1	v_2	v_3	v_4	v_5
v_0	0	(6,6)	0	(10,8)	0	0
v_1	0	0	(3,3)	0	(5 ,5)	0
v_2	0	0	0	0	(4,0)	(9,5)
v_3	0	(3,2)	(2,2)	0	(5,4)	0
v_4	0	0	0	0	0	(9,9)
v_5	0	0	0	0	0	0

Table 8.3 Net input flow table for Example 8.5

Vertex v_i	Net input flow $\phi(v_i)$
v_0	14
v_1	0
v_2	0
v_3	0
v_4	0
v_5	-14

Note that indeed the principle of conservation of flow as ensured via the conclusion of Theorem 8.4 is indeed satisfied

$$\sum_j \phi(v_j) = 0.$$

8.3 Formulation of Maximal Flow Network Problem

Given a capacitated directed network $N = (V, E)$ with unique fixed source v_s, unique fixed destination v_d and having no edges into the source and no edges out of the sink, the first network problem that we consider is to find the *maximal flow* through the network.

Definition 8.6 (Maximal flow network problem)**.** Let $N = (V, E)$ be a capacitated directed network with unique fixed source v_s and unique fixed destination v_d having no edges into v_s and no edges out of v_d. Then the corresponding *maximal flow network problem* is formulated as

- Maximize the objective function

$$f = \sum_i x_{id}$$

- Subject to the constraints

$$\sum_i x_{ij} - \sum_i x_{ji} = 0 \text{ for all } v_j \in V,\ j \neq s, d$$

and

$$0 \leq x_{ij} \leq c_{ij} \text{ for all } i,\ j.$$

(In each of the items above, we consider only those decision variables $x_{\ell k}$ which are such that $(v_\ell, v_k) \in E$.)

Maximal flow network problems can be solved using variants of the Simplex Method; however they can also be solved using more specialized graph theoretic techniques. Here we focus on solving maximal flow network problems via the Simplex Method.

Example 8.7. As a small variant to Example 8.1, consider a capacitated network described in Figure 8.1 together with a flow with $v_s = v_0$ and $v_d = v_5$.

Here we modify slightly the flow capacities c_{ij} from Example 8.1. They are given in the following *capacity flow constraint table* (see Table 8.4).

In this table, the i, j entry c_{ij} is the edge weight from the edge v_i to the edge v_j. In particular $c_{ij} = 0$ if there is no edge from vertex v_i to vertex v_j and we view c_{ij} as representing the capacity of the directed edge from v_i to v_j.

The corresponding maximal network flow problem is described as

- Maximize the objective function

$$f = x_{25} + x_{45}$$

- Subject to the constraints

$$\begin{aligned}
x_{01} + x_{31} - x_{12} - x_{14} &= 0 \\
x_{12} + x_{32} - x_{25} - x_{24} &= 0 \\
x_{03} - x_{31} - x_{32} - x_{34} &= 0 \\
x_{14} + x_{34} + x_{24} - x_{45} &= 0
\end{aligned}$$

Table 8.4 Capacity flow constraint table for Example 8.7

	v_0	v_1	v_2	v_3	v_4	v_5
v_0	0	8	0	9	0	0
v_1	0	0	3	0	5	0
v_2	0	0	0	0	4	10
v_3	0	3	2	0	5	0
v_4	0	0	0	0	0	9
v_5	0	0	0	0	0	0

Table 8.5 Network flow capacity table for Example 8.7

	v_0	v_1	v_2	v_3	v_4	v_5
v_0	0	(8,8)	0	(9,6)	0	0
v_1	0	0	(3,3)	0	(5 ,5)	0
v_2	0	0	0	0	(4,0)	(10,5)
v_3	0	(3,0)	(2,2)	0	(5,4)	0
v_4	0	0	0	0	0	(9,9)
v_5	0	0	0	0	0	0

$$x_{01} \leq 8, x_{03} \leq 9, x_{12} \leq 3, x_{14} \leq 5, x_{24} \leq 4, x_{25} \leq 10$$

$$x_{31} \leq 3, x_{32} \leq 2, x_{34} \leq 5, x_{45} \leq 9 \text{ and}$$

$$x_{ij} \geq 0, \text{for all } i, j.$$

Using the Simplex Method a maximal flow is found to be

$$f = 14$$

which is achieved when the decision variables x_{ij} take on the values that are described in the network flow capacity Table 8.5 which has i, j entry described by (c_{ij}, x_{ij}).

8.4 Matrix Representation of Weighted Digraphs

In this section, we develop matrix notation to encode the data of weighted directed networks. This approach is very natural from the vantage point of algebraic graph theory and especially matrix approaches for working with *weighted digraphs*.

The *adjacency matrix* of a digraph

$$N = (V, E)$$

with vertex set

$$V = \{v_1, \ldots, v_n\}$$

is the $n \times n$ binary matrix

$$R = [r_{ij}]$$

where

$$r_{ij} = \begin{cases} 1 & \text{if there is a directed edge from } v_i \text{ to } v_j \\ 0 & \text{otherwise.} \end{cases}$$

The *weight adjacency matrix* of a weighted digraph

$$N = (V, E)$$

with vertex set

$$V = \{v_1, \dots, v_n\}$$

is the $n \times n$ matrix

$$C = [c_{ij}]$$

where c_{ij} is the weight, or capacity, that is associated to the edge

$$(v_i, v_j) \in E.$$

(So, $c_{ij} = 0$ if there is no edge from v_i to v_j.)

For our purposes, at times we use interchangeably the terminology *weighted digraph with no multiple directed edges* and *capacitated directed network*. In other words, the capacitated directed networks that we consider here are *simple weighted digraphs*.

Example 8.8. The network from Example 8.7 has adjacency matrix given by

$$R = \begin{bmatrix} 0 & 1 & 0 & 1 & 0 & 0 \\ 0 & 0 & 1 & 0 & 1 & 0 \\ 0 & 0 & 0 & 0 & 1 & 1 \\ 0 & 1 & 1 & 0 & 1 & 0 \\ 0 & 0 & 0 & 0 & 0 & 1 \\ 0 & 0 & 0 & 0 & 0 & 0 \end{bmatrix}$$

and weight adjacency matrix given by

$$C = \begin{bmatrix} 0 & 8 & 0 & 9 & 0 & 0 \\ 0 & 0 & 3 & 0 & 5 & 0 \\ 0 & 0 & 0 & 0 & 4 & 10 \\ 0 & 3 & 2 & 0 & 5 & 0 \\ 0 & 0 & 0 & 0 & 0 & 9 \\ 0 & 0 & 0 & 0 & 0 & 0 \end{bmatrix}.$$

8.5 Formulation of Shortest Path Network Problem

In this section, we give a formulation of the *shortest path network problem.*

For this purpose, let

$$N = (V, E)$$

be a *directed network*. Then N is said to be *weighted* if to each edge

$$(v_i, v_j) \in E$$

there corresponds a real number w_{ij} (not necessarily nonnegative) called its *weight*. The data of a weighted directed network is a generalization of a capacitated directed network.

Let

$$N = (V, E)$$

be a directed network. A finite nonempty set P of edges of E of the form

$$P = \{(v_0, v_1), (v_1, v_2), \ldots, (v_{n-1}, v_n)\}$$

is said to be a *path* in N from v_0 to v_n. If $v_0 = v_n$, then the path is a *cycle*.

The intuitive formulation of the *shortest path network problem* is actually quite elementary. The idea is to consider a *weighted directed network* N and then to interpret the weights of the edges as distances between corresponding vertices.

The *network* can be viewed as a *map* with *vertices representing cities* and *directed edges representing one-way roads* between these cities.

The *shortest path network problem* is then concerned with *finding the path of shortest distance* from one distinguished vertex, called the *derivation*, or *source*, to another distinguished vertex called the *destination*.

A precise formulation of the *shortest path network problem* is achieved in the following way.

Definition 8.9 (Shortest path network problem)**.** Let

$$N = (V, E)$$

be a weighted directed network with no edges into the derivation v_s and no edges out of the destination v_d. Further, assume that no cycle in N has a net negative weight—in the sense that no cycle in N consists of edges the sum of whose weights is negative.

In this context, the *shortest path network problem* seeks to

- Minimize the linear objective function

$$\text{dist} = \sum_{(v_i, v_j) \in E} w_{ij} x_{ij}$$

- Subject to the constraints

$$\sum_i x_{id} = 1$$

$$\sum_i x_{ij} - \sum_i x_{ji} = 0 \text{ for all } v_j \in V \text{ and } j \neq s, d$$

and

$$0 \leq x_{ij} \leq 1 \text{ for all } i, j \text{ with } (v_i, v_j) \in E.$$

(In each of the items above, we only consider decision variables $x_{\ell k}$ with $(v_\ell, v_k) \in E$.)

Remark 8.10. In our formulation of the shortest path network problem, see Definition 8.9, if we did not require that there are no edges into the derivation v_s, then we would need to add the additional constraint that $\sum_j x_{sj} = 1$.

It turns out that there is a graph theoretic algorithm that is due to *Dijkstra* which solves the shortest path network problem. On the other hand, the shortest path network problem can also be solved using the Simplex Method and its variants. A (non-obvious) feature of such formulations of the shortest path network problem as a linear programming problem is that in each optimal solution the decision variables will take on values of either 0 or 1.

Example 8.11. Consider the weighted directed network which has *weight adjacency table* described in Table 8.6.

In Table 8.6, the i, j entry w_{ij} is the weight that is assigned to the directed edge $(v_i, v_j) \in E$. A graphical representation of the directed network is described in Figure 8.2.

Table 8.6 Weight adjacency table for Example 8.11

	v_0	v_1	v_2	v_3	v_4	v_5
v_0	0	3	0	2	0	0
v_1	0	0	6	3	0	0
v_2	0	1	0	2	0	3
v_3	0	2	0	0	4	0
v_4	0	0	1	3	0	5
v_5	0	0	0	0	0	0

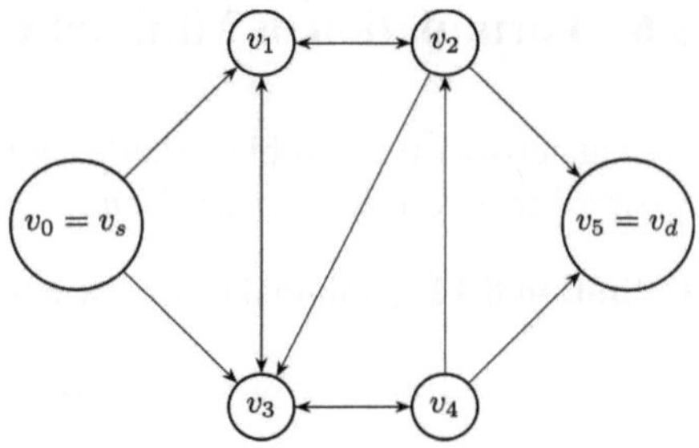

Fig. 8.2 Graphical representation of the directed network for Example 8.11

The corresponding shortest path network problem can then be formulated as

- Minimize the objective function

$$\text{dist} = \sum_{(v_i, v_j) \in E} w_{ij} x_{ij} = 3x_{01} + 2x_{03} + 6x_{12} + 3x_{13} + x_{21} + 2x_{23} + 3x_{25} + 2x_{31} + 4x_{34} + x_{42} + 3x_{43} + 5x_{45}$$

- Subject to the constraints

$$x_{25} + x_{45} = 1$$

$$x_{01} + x_{21} + x_{31} - x_{12} - x_{13} = 0$$

$$x_{12} + x_{42} - x_{21} - x_{23} - x_{25} = 0$$

$$x_{03} + x_{13} + x_{23} + x_{43} - x_{34} - x_{31} = 0$$

$$x_{34} - x_{42} - x_{45} = 0$$

and

$$0 \leq x_{ij} \leq 1 \text{ for all } i, j \text{ with } (v_i, v_j) \in E.$$

Using the Simplex Method, an optimal solution is achieved when $x_{03} = x_{34} = x_{42} = x_{25} = 1$ and all other $x_{ij} = 0$ in which case

$$\text{dist} = 10.$$

8.6 Formulation of Minimal Cost Flow Network Problem

A final network type problem that we will consider, for the present time being, is the *minimal cost flow network problem.*

Definition 8.12 (Minimal cost flow network problem)**.** Let

$$N = (V, E)$$

be a capacitated directed network with unique fixed source v_s and unique fixed sink v_d and having no edges into the source and having no edges out of the sink.

Assume that to each edge

$$(v_i, v_j) \in E$$

there also corresponds a real number c'_{ij} which is to be interpreted as a *cost*.

The *minimal cost flow network problem* is formulated as

- Minimize the objective function

$$C = \sum_{(v_i,v_j)\in E} c'_{ij} x_{ij}$$

- Subject to the constraints

$$\sum_i x_{id} = F \geq 0$$

$$\sum_i x_{ij} - \sum_i x_{ji} = 0 \text{ for all } v_j \in V,\ j \neq s, d$$

and

$$0 \leq x_{ij} \leq c_{ij} \text{ for all } i,\ j.$$

Here, F is a given specified flow value. (In each of the above summations, we only consider decision variables $x_{\ell k}$ with $(v_\ell, v_k) \in E$.)

Remark 8.13. In our formulation of the minimal cost flow network problem, as in Definition 8.12, in case that there are multiple sinks $v_{d_1}, \ldots, v_{d_n}$ with specified flow values $F_1, \ldots, F_n \geq 0$ we then consider the constraints $\sum_i x_{id_\ell} = F_\ell \geq 0$, for $\ell = 1, \ldots, n$ in place of the single flow constraint $\sum_i x_{id}$.

The *minimal cost flow network problem* can be solved by the Simplex Method and/or more specialized variants thereof.

Example 8.14. Here, we consider the weighted directed network that is determined by the following weight adjacency table (see Table 8.7).

Table 8.7 Weight adjacency table for Example 8.14

	v_0	v_1	v_2	v_3	v_4	v_5
v_0	0	(7,1)	0	(5,2)	0	0
v_1	0	0	(5,3)	0	(5,2)	0
v_2	0	0	0	0	(3,2)	(5,1)
v_3	0	(2,6)	0	0	(5,3)	0
v_4	0	0	0	0	0	(8,1)
v_5	0	0	0	0	0	0

In the weight adjacency table, the (nonzero) label that is associated to each edge should be interpreted as

$$(c_{ij}, c'_{ij}) = (c_{ij}, c'_{ij}).$$

In other words, the first component is the capacity of the edge and the second component is the cost of the edge. A zero entry means that there is no edge from vertex v_i to vertex v_j.

The corresponding minimal cost flow network problem with given specified flow value

$$F = 10$$

is formulated as follows.

- Minimize the linear cost function

$$C = x_{01} + 2x_{03} + 3x_{12} + 2x_{14} + 2x_{24} + x_{25} + 6x_{31} + 3x_{34} + x_{45}$$

- Subject to the constraints

$$x_{25} + x_{45} \geq 10$$

$$x_{01} + x_{31} - x_{12} - x_{14} = 0$$

$$x_{12} - x_{24} - x_{25} = 0$$

$$x_{03} - x_{31} - x_{34} = 0$$

$$x_{14} + x_{24} + x_{34} - x_{45} = 0$$

and

$$0 \leq x_{01} \leq 7, 0 \leq x_{03} \leq 5, 0 \leq x_{12} \leq 5,$$
$$0 \leq x_{14} \leq 5, 0 \leq x_{24} \leq 3, 0 \leq x_{25} \leq 5,$$
$$0 \leq x_{31} \leq 6, 0 \leq x_{34} \leq 3, 0 \leq x_{45} \leq 8.$$

Using the Simplex Method, an optimal solution is found to be

$$C = 48.$$

This is achieved when $x_{01} = 7, x_{03} = 3, x_{34} = 3, x_{12} = 2, x_{14} = 5, x_{45} = 8, x_{25} = 2.$

Exercises

8.1 Consider the weighted directed network that is determined by the weight adjacency table given in Table 8.8.

(i) Give a graphical description of the corresponding network.
(ii) Interpret the arrows as one-way routes between the nodes and interpret the numbers as the distance between the nodes.
(iii) Find the shortest distance from node v_0 to node v_5 as well as an actual route from node v_0 to node v_5 which achieves this shortest distance.

8.2 Consider the capacitated directed network that is determined by the network flow capacity table given in Table 8.9.

A zero entry, in Table 8.9, means that there is no edge from vertex v_i to vertex v_j.

(i) Give a graphical description of the corresponding network.
(ii) Solve the corresponding minimal cost flow network problem with $F = 9$.

Table 8.8 Weight adjacency table for Exercise 8.1

	v_0	v_1	v_2	v_3	v_4	v_5
v_0	0	4	0	2	0	0
v_1	0	0	3	0	2	0
v_2	0	0	0	1	3	0
v_3	0	1	0	0	4	0
v_4	0	0	0	0	0	3
v_5	0	0	0	0	0	0

Table 8.9 Flow capacity table for Exercise 8.2

	v_0	v_1	v_2	v_3	v_4
v_0	0	(3,4)	(5,2)	(5,5)	0
v_1	0	0	(2,3)	0	0
v_2	0	0	0	(3,2)	(7,5)
v_3	0	0	0	0	(6,2)
v_4	0	0	0	0	0

Table 8.10 Transportation tableau for Exercise 8.3

	M_1	M_2	M_3	
W_1	2	1	2	40
W_2	2	1	2	40
W_3	1	2	9	100
	40	50	20	

Table 8.11 Network flow capacity table for Exercise 8.4

	v_1	v_2	v_3	v_4	v_5	v_6	v_7
v_1	0	(4,1)	0	0	(5,0)	0	0
v_2	0	0	(8,6)	0	0	0	0
v_3	0	(4,2)	0	(6,0)	(5,5)	0	0
v_4	0	0	0	0	0	0	(8,4)
v_5	0	(2,1)	0	(7,4)	0	0	0
v_6	(6,5)	0	0	0	0	0	(7,1)
v_7	0	0	0	0	(3,1)	(4,3)	0

8.3 Reformulate the following unbalanced transportation problem, with transportation tableau given in Table 8.10, as a minimal cost flow network problem.

8.4 Consider the capacitated directed network $N = (V, E)$ with flow that is specified by the following network flow capacity table. (See Table 8.11.)

In Table 8.11, the label that is associated to each edge should be interpreted as (c_{ij}, x_{ij}) where c_{ij} is the i, j-capacity and x_{ij} is the i, j-flow. A zero entry means that there is no edge from vertex v_i to vertex v_j.

(i) Give a graphical description of the corresponding network.
(ii) For each vertex $v_j \in V$, compute the quantity $\phi(v_j)$. Verify the conservation of flow in N.
(iii) Classify each vertex $v_j \in V$ as a source, a sink, or an intermediate vertex.

8.5 Consider the capacitated directed network $N = (V, E)$ with flow that is specified by the weight adjacency matrix which is given in Table 8.12.

Table 8.12 Weight adjacency matrix for Exercise 8.5

	v_0	v_1	v_2	v_3	v_4	v_5	v_6
v_0	0	4	0	3	2	0	0
v_1	0	0	1	1	0	0	0
v_2	0	0	0	0	0	0	3
v_3	0	0	1	0	0	1	1
v_4	0	0	0	1	0	2	0
v_5	0	0	0	0	0	0	4
v_6	0	0	0	0	0	0	0

In Table 8.12, the (nonzero) label that is associated to each edge should be interpreted as the i, j-capacity c_{ij}. A zero entry means that there is no edge from vertex v_i to vertex v_j.

(i) Give a graphical description of the corresponding network.
(ii) Solve the corresponding maximal flow network problem.

Chapter 9
Selected Introduction to Nonlinear Programming

Overview

- Introduction to nonlinear programming problems.
- Classical multivariable unconstrained optimization.
- Convex programming and Karush-Kuhn-Tucker conditions.
- Quadratic programming problems.
- Separable programming problems.

9.1 Formulation of Nonlinear Programming Problems

A general formulation of a *nonlinear maximization programming problem* is

- Maximize the objective function

$$f(x_1, \dots, x_n)$$

- Subject to the constraints

$$g_i(x_1, \dots, x_n) \leq b_i \text{ for } i = 1, \dots, m$$

and

$$x_i \geq 0 \text{ for } i = 1, \dots, n.$$

Here the objective function $f(x_1, \dots, x_n)$ and constraint functions $g_i(x_1, \dots, x_n)$ are, à priori, *nonlinear functions* in the decision variables $x_1, \dots, x_n$. *Nonlinear minimization problems* may be formulated similarly. In formulating such nonlinear maximization and/or minimization programming problems constraints may include equalities $=$ in addition to inequalities such as $\geq$ and/or $\leq$.

N. Grieve, *An Introduction to Mathematical Programming and Network Science*,
Springer Undergraduate Texts in Mathematics and Technology,
https://doi.org/10.1007/978-3-032-13330-4_9

There are many different types of nonlinear programming problems. An important class of nonlinear programming problems arise when the objective function and the constraint functions exhibit various features of convexity. In what follows, we survey some techniques for solving convex nonlinear programming problems. However, before doing so, we give a brief overview of some motivational applications.

Example 9.1. (Product Mixing Problems with Price Elasticity). In *product mixing problems*, the goal is to determine the optimal mix of production levels for a manufacture's products, given limitations on the resources needed to produce those products, in order to maximize the manufacture's total profit.

In some cases, there is a fixed unit profit that is associated with each product—so the resulting objective function will be linear. However, in many product mixing problems, there may be *price elasticity* whereby the amount of a product that can be sold has an inverse relationship to the price charged.

For example, there might be a *price demand curve* $p(x)$ as pictured in Figure 9.1.

Thus, if the unit cost for production and distribution of the product is fixed at c then the manufacturing firm's profit from producing and selling x units is given by the nonlinear function

$$P(x) = xp(x) - cx.$$

The graph of such a function is pictured as in Figure 9.2.

More generally, if each of the firm's n products has profit function

$$P_j(x_j) = x_j p(x_j) - c_j x_j$$

where the unit cost for production and distribution of the jth product is fixed at c_j then the overall profit function would take the form

$$P(x_1, \ldots, x_n) = \sum_{j=1}^{n} P_j(x_j).$$

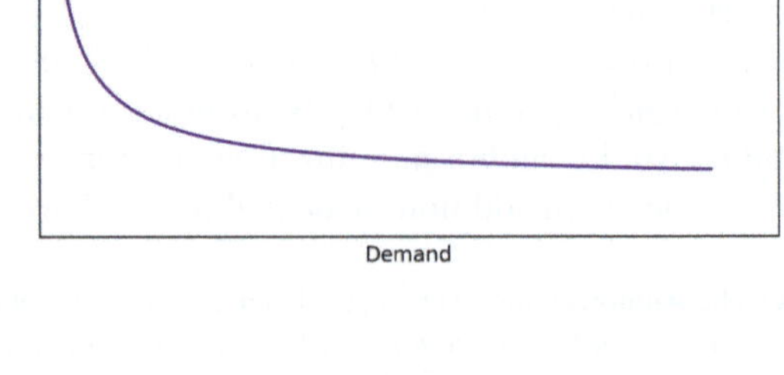

Fig. 9.1 Price demand curve for Example 9.1

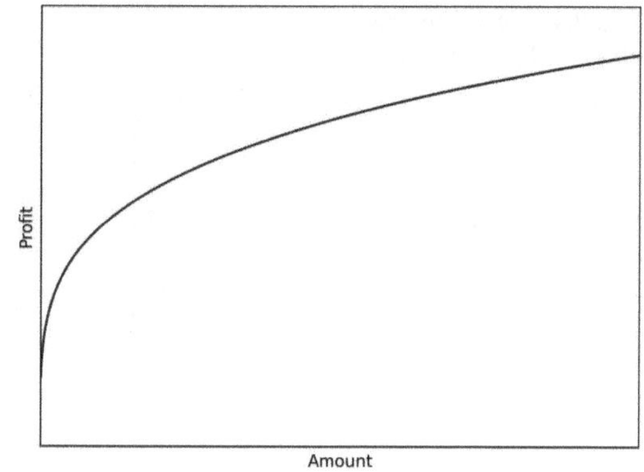

Fig. 9.2 Profit function for Example 9.1

Example 9.2. (Transportation Problem with Volume Discounts on Shipping Costs). Recall, that a typical application of the *transportation problem* is to determine an optimal plan for shipping goods from various sources to various destinations, given supply and demand constraints in order to minimize total shipping cost.

In the transportation problems that we have considered in Section 7.2, we have always assumed that the cost per unit shipped from a given source to a given destination is fixed, regardless of the amount shipped. We now want to allow for more general models. Indeed, certain *volume discounts* might be available for large shipments.

So, for example, the *marginal cost* of shipping might be described graphically as in Figure 9.3.

The resulting cost of shipping x units is then given by a *nonlinear function* $C(x)$ which is actually a *piecewise linear function*. The slope of each linear component equals the corresponding marginal cost. The total shipping cost is obtained via integration of the marginal shipping cost.

This is described graphically in Figure 9.4.

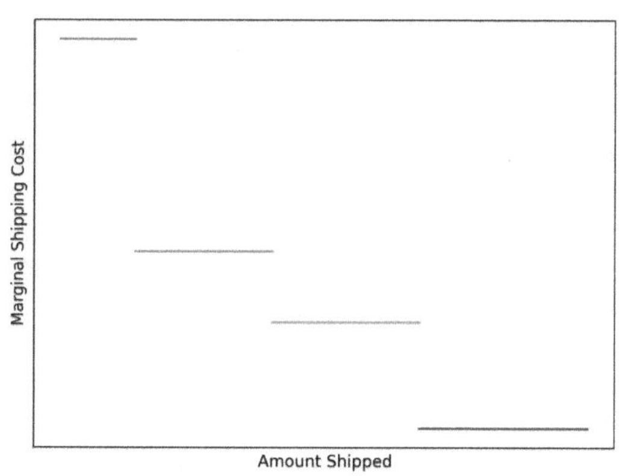

Fig. 9.3 Marginal shipping costs for Example 9.2

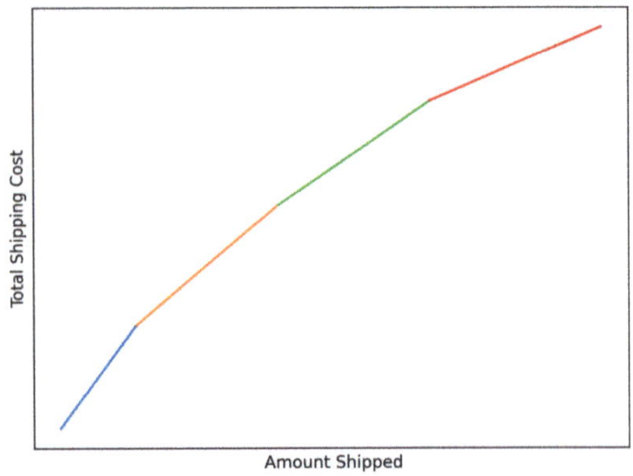

Fig. 9.4 Total shipping costs for Example 9.2

Generally speaking, if each combination of source and destination has a *particular shipping cost function*, so, for example, the cost of shipping x_{ij} units from source i, $i = 1, \ldots, m$ to destination j, $j = 1, \ldots, n$ is given by a nonlinear function $C_{ij}(x_{ij})$, then the overall objective function to be *minimized* is given by

$$f(x_{11}, \ldots, x_{mn}) = \sum_{i=1}^{m} \sum_{j=1}^{n} C_{ij}(x_{ij}).$$

In practice, with such nonlinear objective functions, the constraints for such *transportation problems with volume discounts* may take a more specialized nonlinear form.

In Example 9.3, we consider another motivational class of examples. They arise from the question of borrowing given variable interest rates.

Example 9.3. (Borrowing with variable interest rates)**.** A developer is concerned both with the *expected cost* (interest paid) and the *volatility* (variability of interest rate) that is associated with their financing products. To assist with this, a certain (nonlinear) *quadratic programming model* can be used.

For example, suppose that n lending options are being considered for inclusion in the financing plan and let the decision variables x_j, $j = 1, \ldots, n$, be the amount borrowed from financing option j.

Let μ_j and σ_{jj}, respectively, be the estimated *mean* and *variance* of the interest rate on borrowing option j where σ_{jj} measures the variability of the interest rate of option j.

For $i = 1, \ldots, n$, $i \neq j$, let σ_{ij} be the *covariance* of the return on one share of each of stock i and stock j. It can be difficult to estimate covariance values. However, with suitable hypothesis, estimated values for σ_{ij} can be determined from σ_{ii} and σ_{jj}. We refer to Section 9.7.1 for one approach to this.

Respectively, the expected interest cost $I(x_1, \ldots, x_n)$ and the variance, that is the interest rate variability, $V(x_1, \ldots, x_n)$ of the financing portfolio's total borrowing cost are given by

$$I(x_1, \ldots, x_n) = \sum_{j=1}^{n} \mu_j x_j$$

and

$$V(x_1, \ldots, x_n) = \sum_{i=1}^{n} \sum_{j=1}^{n} \sigma_{ij} x_i x_j.$$

Note that $V(x_1, \ldots, x_n)$ gives a measure of the *interest rate variability* that is associated with the financing portfolio.

Let L be the maximum expected borrowing cost and for each $j = 1, \ldots, n$, let I_j be the estimated interest rate for the jth borrowing option. Let B be the amount of money that is budgeted for required capital and interest payments.

Then, in this context, the *borrowing with variable interest rates* programming problem is formulated in the following way.

- Minimize the objective function

$$V(x_1, \ldots, x_n) = \sum_{i=1}^{n} \sum_{j=1}^{n} \sigma_{ij} x_i x_j$$

- Subject to the constraints

$$\sum_{j=1}^{n} \mu_j x_j \leq L$$

$$\sum_{j=1}^{n} I_j x_j \geq B$$

and

$$x_j \geq 0 \text{ for } j = 1, \ldots, n.$$

This borrowing with variable interest rates quadratic programming problem that we have formulated here is complementary to the *Portfolio Selection Problem* which are formulated in Section 9.7.

9.2 Selected Techniques for Single Variable Programming Problems

In this section, we describe a collection of classical techniques for studying single variable programming problems.

A one-variable *unconstrained* nonlinear program has the form

- Maximize and/or Minimize the objective function

$$Z = f(x)$$

for $f(x)$ a (nonlinear) single variable function of a single variable x and search for the optimum (maximum or minimum) for $-\infty < x < \infty$.

A one-variable *constrained* nonlinear program has the form

- Maximize and/or Minimize the objective function

$$Z = f(x) \text{ for } a \leq x \leq b.$$

Solving such single variable unconstrained and constrained optimization problems, provide applications of the first and second derivative test along with the closed interval method from single variable calculus.

Another classical method from single variable calculus is *Newton's Method*. The idea is to approximate $f(x)$ within the neighbourhood of the current trial solution by a suitable quadratic function and then to maximize (or minimize) the approximate function exactly in order to obtain a new trial solution to start the next iteration.

It turns out that the approximating quadratic function is obtained by truncating the Taylor series after the second derivative term. In more detail, letting x_{i+1} be the trial solution generated at iteration i, to start iteration $i+1$, the *truncated Taylor series* for x_{i+1} is

$$f(x_{i+1}) \approx f(x_i) + f'(x_i)(x_{i+1} - x_i) + \frac{f''(x_i)}{2}(x_{i+1} - x_i)^2.$$

This quadratic function can be maximized by setting its first derivative to zero and then solving for x_{i+1}.

The first derivative is

$$f'(x_{i+1}) \approx f'(x_i) + f''(x_i)(x_{i+1} - x_i)$$

(recall x_i, $f(x_i)$ and $f''(x_i)$ are constants). Setting this first derivative to zero yields

$$f'(x_i) + f''(x_i)(x_{i+1} - x_i) = 0;$$

upon solving for x_{i+1} it then follows that

$$x_{i+1} := x_i - \frac{f'(x_i)}{f''(x_i)}.$$

This is the key formula that gets used at each iteration of Newton's Method to calculate the next trial solution.

A concise formulation of Newton's Method to optimize $f(x)$ is achieved in the following way. In our formulation, we note that Newton's Method does not always optimize a given function. For instance, a reasonable choice of trial solution and error tolerance are required as input.

Newton's Method

- **Input.** Error tolerance $\epsilon > 0$, initial trial solution x_1, set $i = 1$.
- **Iteration** i**:**
  - Calculate $f'(x_i)$ and $f''(x_i)$.
  - Set $x_{i+1} := x_i - \frac{f'(x_i)}{f''(x_i)}$.
- **Stopping Rule.** If $|x_{i+1} - x_i| \leq \epsilon$, then stop; x_{i+1} is essentially the optimal solution. Otherwise reset $i := i + 1$ and perform another iteration.

Example 9.4. Consider the case of Newton's Method applied to

$$f(x) = 12x - 3x^4 - 2x^6.$$

The key iterative formula takes the form

$$x_{i+1} := x_i - \frac{f'(x_i)}{f''(x_i)} = x_i + \frac{1 - x_i^3 - x_i^5}{3x_i^2 + 5x_i^4}.$$

The calculation with error tolerance $\epsilon = 0.00001$ and initial value $x_1 = 1$ are summarized in Table 9.1.

Table 9.1 Table of values for Example 9.4

Iteration i	x_i	$f(x_i)$
1	1	7
2	.875	7.8439
3	.84003	7.8838
4	.83763	7.8839

9.3 Graphical Visualization of Nonlinear Programming Problems

When a nonlinear programming problem has just one or two variables, it can be represented graphically. Because such graphical representations give considerable insight into the properties of optimal solutions for linear and nonlinear programming, it is helpful to look into some examples.

First recall from multivariable calculus the manner in which contour diagrams give qualitative features and approximate location of local extrema for functions of two variables. Similar considerations apply to two variable nonlinear optimization problems. However, there are many additional subtle points to be aware of.

Example 9.5. Figure 9.5 depicts an example of a linear objective function together with the presence of a nonlinear constraint function.

Here, in Figure 9.5, the optimal solution is identified as the point of intersection of the linear objective function with the boundary of the feasible region.

Example 9.6. Figure 9.6 gives an example of the contours of a nonlinear objective function together with the presence of only linear constraint functions.

In Figure 9.6, the optimal solution(s) are identified as the maximum and/or minimum value(s) that correspond to the appropriate contours that intersect nontrivially with the feasible region.

Fig. 9.5 Objective function and constraints for Example 9.5

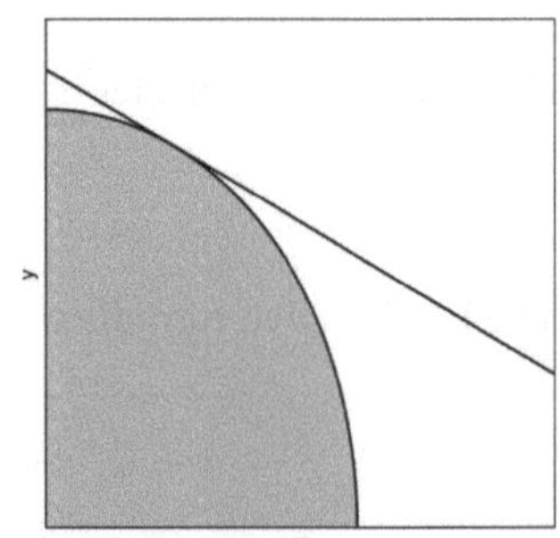

Fig. 9.6 Objective function and constraints for Example 9.6

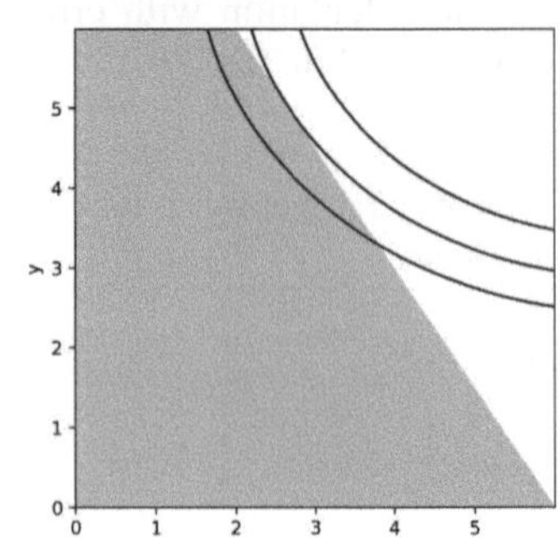

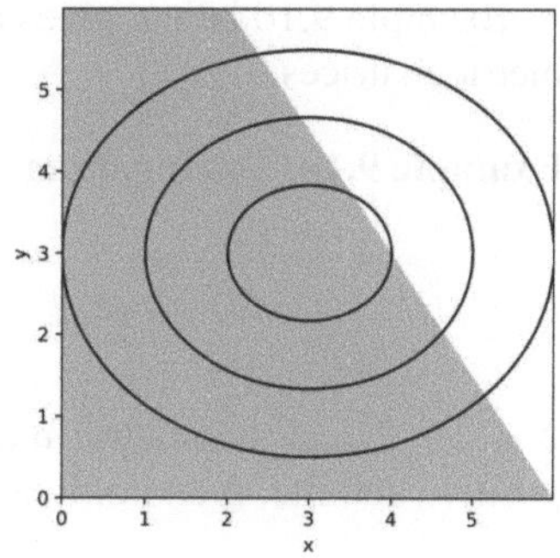

Fig. 9.7 Objective function and contours for Example 9.7

Example 9.7. Figure 9.7 gives an example of the contours of a nonlinear objective function together with the presence of only linear constraint functions.

Here, as in Example 9.6, the optimal solution(s) are identified as the maximum and/or minimum values that correspond to the appropriate contours that intersect nontrivially with the feasible region.

9.4 Classical Multivariable Unconstrained Optimization

The starting point to classical multivariable unconstrained optimization is to generalize to several variables the concept of concave up and concave down for functions of a single variable. This is largely achieved via the theory of positive semi-definiteness for real symmetric matrices. The main points are summarized in Section 9.4.1.

9.4.1 Positive Semi-definite Matrices and Convexity

Definition 9.8. An $n \times n$ real symmetric matrix A is called *positive semi-definite*, respectively, *positive definite*, if

$$\begin{bmatrix} x_1 \dots x_n \end{bmatrix} A \begin{bmatrix} x_1 \\ \vdots \\ x_n \end{bmatrix} \geq 0, \text{ respectively } > 0, \text{ for all } \mathbf{x} = \begin{bmatrix} x_1 \\ \vdots \\ x_n \end{bmatrix} \in \mathbb{R}^n.$$

It turns out that real symmetric matrices are characterized in the following way.

Theorem 9.9. *Let A be an $n \times n$ real symmetric matrix. Then the following assertions are equivalent.*

(i) A is positive semi-definite, respectively, positive definite.
(ii) If $\lambda_1, \dots, \lambda_n$ are the eigenvalues of A, then $\lambda_1 \geq 0, \dots, \lambda_n \geq 0$, respectively, $>$.

Example 9.10 below gives a special case of Theorem 9.9 applied to 2×2 symmetric matrices.

Example 9.10. An important special case of Theorem 9.9 is when

$$A = \begin{bmatrix} a & b \\ b & c \end{bmatrix}$$

is a 2×2 real symmetric matrix. Then A is positive semi-definite, respectively, positive definite, if and only if

$$a \geq 0, c \geq 0 \text{ and } ac - b^2 \geq 0, \text{ respectively } >.$$

We now discuss the concept of convex and strictly convex functions. A key point is the manner in which they can be characterized, respectively, in terms of positive semi-definiteness and positive definiteness of *Hessian matrices.*

In fact, in Definition 9.11, we actually use this characterization as our starting point in defining *convexity*, *strict convexity*, *concavity* and *strict concavity* for several variable functions.

Definition 9.11. Let $f(x_1, \ldots, x_n)$ be a twice differentiable real-valued function on an open convex subset $C \subseteq \mathbb{R}^n$.

- $f(x_1, \ldots, x_n)$ is *convex*, respectively, *strictly convex*, on C if and only if its *Hessian matrix*
$$\text{Hess}(f) := \left[\frac{\partial^2 f}{\partial x_i \partial x_j}(x_1, \ldots, x_n)\right]$$
is positive semi-definite, respectively, positive definite at all $\mathbf{x} = \mathbf{x}_0 \in C$.
- $f(x_1, \ldots, x_n)$ is *concave*, respectively, *strictly concave*, on C if $-f(x_1, \ldots, x_n)$ is convex, respectively, strictly convex at all $\mathbf{x} = \mathbf{x}_0 \in C$.

Example 9.12. When $n = 2$, the condition that the Hessian matrix of a twice differentiable function $f(x_1, x_2)$ on $\mathbb{R}^n$ is convex amounts to the three conditions that

$$f_{x_1x_1} f_{x_2x_2} - f_{x_1x_2}^2 \geq 0$$

$$f_{x_1x_1} \geq 0$$

and

$$f_{x_2x_2} \geq 0.$$

Here is a concrete instance of Example 9.12.

Example 9.13. If

$$f(x_1, x_2) = (x_1 - x_2)^2 = x_1^2 - 2x_1x_2 + x_2^2$$

then $f(x_1, x_2)$ is convex since

$$f_{x_1x_1} f_{x_2x_2} - f_{x_1x_2}^2 = (2)(2) - (-2)^2 = 0$$

$$f_{x_1x_1} = 2 > 0$$

and

$$f_{x_2x_2} = 2 > 0.$$

However, $f(x_1, x_2)$ is not strictly convex since not all of these three quantities are positive.

In Theorems 9.14, 9.15 and 9.16 below, we give a summary of the main results for unconstrained optimization of several variable functions. These results should be seen as an extension of the main result about optimization to bi-variable functions which is recalled in Appendix B.

Theorem 9.14. *Let $f(x_1, \ldots, x_n)$ be a continuous function defined on some closed and bounded region of $\mathbb{R}^n$. Then $f(x_1, \ldots, x_n)$ has a global maximum and a global minimum on that region.*

Theorem 9.15. *If $f(x_1, \ldots, x_n)$ has a local maximum or a local minimum at some point $\mathbf{x} = \mathbf{x}_0 \in \mathbb{R}^n$ and if its gradient vector*

$$\nabla f := \langle \partial f/\partial x_1, \ldots, \partial f/\partial x_n \rangle$$

exists on some open ball with centre $\mathbf{x} = \mathbf{x}_0$, then

$$\nabla f|_{\mathbf{x}=\mathbf{x}_0} = \mathbf{0}.$$

Theorem 9.16. *If $f(x_1, \ldots, x_n)$ has second partial derivatives on some open ball with centre $\mathbf{x} = \mathbf{x}_0 \in \mathbb{R}^n$, if*

$$\nabla f|_{\mathbf{x}=\mathbf{x}_0} = \mathbf{0}$$

and if

$$-\operatorname{Hess}(f)|_{\mathbf{x}=\mathbf{x}_0} = -\left[\frac{\partial^2 f}{\partial x_i \partial x_j}(x_1, \ldots, x_n)\right]\Bigg|_{\mathbf{x}=\mathbf{x}_0}$$

is positive definite, then $\mathbf{x} = \mathbf{x}_0$ is a local maximum.

Furthermore, if

$$\text{Hess}(f)|_{\mathbf{x}=\mathbf{x}_0} = \left[\frac{\partial^2 f}{\partial x_i \partial x_j}(x_1, \ldots, x_n)\right]\Bigg|_{\mathbf{x}=\mathbf{x}_0}$$

is positive definite for $\mathbf{x} = \mathbf{x}_0 \in \mathbb{R}$, *then* $\mathbf{x} = \mathbf{x}_0$ *is a local minimum.*

9.5 The Methods of Steepest Descent and Lagrange Multipliers

We want to describe in further details some of the main classical methods for multivariable optimization. First, in Section 9.5.1, we describe the method of Steepest Descent for non-constrained programming problems. Then we recall, in Section 9.5.2, some aspects to the method Lagrange Multipliers which gives an approach to solving certain multivariable optimization problems with equality constraints.

9.5.1 The Method of Steepest Descent

Suppose that $f(x_1, \ldots, x_n)$ is a twice differentiable function over some region of $\mathbb{R}^n$. Then the *stationary points*, or *critical points*, of $f(x_1, \ldots, x_n)$ are the solutions to

$$\nabla f(x_1, \ldots, x_n) = \mathbf{0}.$$

We want to understand which stationary points give rise to *extreme values* of $f(x_1, \ldots, x_n)$.

For example, recall that $f(x_1, \ldots, x_n)$ has a *local minimum* at

$$\mathbf{x} = \mathbf{x}_0 \in \mathbb{R}^n$$

if

$$f(\mathbf{x}) \geq f(\mathbf{x}_0)$$

for all $\mathbf{x}$ in some neighbourhood of $\mathbf{x}_0$. On the other hand, recall that $f(x_1, \ldots, x_n)$ has a *local maximum* at

$$\mathbf{x} = \mathbf{x}_0 \in \mathbb{R}^n$$

if

$$f(\mathbf{x}) \leq f(\mathbf{x}_0)$$

for all $\mathbf{x}$ in some open neighbourhood of $\mathbf{x}_0$.

The method of *Steepest Descent* provides an iterative method to find local extreme values. The idea, is to find a minimum of $f(\mathbf{x})$ by repeatedly computing minima of a certain function $g(t)$. It is formulated in the following way.

Method of Steepest Descent

- Suppose that $f(\mathbf{x})$ has a minimum at $\mathbf{x}_0$ and that we start at a point $\mathbf{x}$.
- Look for a minimum of $f(x_1, \ldots, x_n)$ closest to $\mathbf{x}$ along the line in the direction of $-\nabla f(\mathbf{x})$; this is the *direction of steepest descent* of $f(x_1, \ldots, x_n)$ at $\mathbf{x}$.
- Determine the value of t and the corresponding point

$$\mathbf{z}(t) = \mathbf{x} - t\nabla f(\mathbf{x})$$

at which the function

$$g(t) := f(\mathbf{z}(t))$$

has a minimum.
- Then, this $\mathbf{z}(t)$ is taken as the next approximation to eventually finding an approximate optimal minimum value of $f(\mathbf{x})$.

Example 9.17. Here, we use the method of Steepest Descent to estimate a minimum of

$$f(x_1, x_2) = x_1^2 + 3x_2^2$$

and starting from $\mathbf{x} = (6, 3)$. (Note that it is not difficult to observe directly that $f(\mathbf{x})$ has a minimum at $\mathbf{x}_0 = (0, 0)$.)

Note that

$$\nabla f(\mathbf{x}) = \langle 2x_1, 6x_2 \rangle.$$

Thus

$$\mathbf{z}(t) = \mathbf{x} - t\nabla f(\mathbf{x}) = \langle (1 - 2t)x_1, (1 - 6t)x_2 \rangle$$

and

$$g(t) = f(\mathbf{z}(t)) = (1 - 2t)^2 x_1^2 + 3(1 - 6t)^2 x_2^2.$$

Therefore

$$g'(t) = 2(1 - 2t)x_1^2(-2) + 6(1 - 6t)x_2^2(-6).$$

Now, setting

$$g'(t) = 0$$

and solving for t yields that

$$t := \frac{x_1^2 + 9x_2^2}{2x_1^2 + 54x_2^2}.$$

Then, starting from

$$\mathbf{x}_0 := \langle 6, 3 \rangle.$$

we compute a table of values as illustrated in Table 9.2.

A graphical picture of the relevant vector fields for

$$-\nabla f(\mathbf{x}) = \langle -2x_1, -6x_2 \rangle$$

along with the contours is given in Figure 9.8 below.

Note that the gradient vector fields $\nabla f(\mathbf{x})$ are outward pointing which means that, as illustrated in the above diagram, that the path chosen by steepest decent in the given minimization problem is in the direction opposite to that which the gradient vector fields are pointing.

Table 9.2 Table of values for Example 9.17

n	$\mathbf{x}$	t
0	$\langle 6.000, 3.000 \rangle$	0.210
1	$\langle 3.484, -0.774 \rangle$	0.310
2	$\langle 1.327, 0.664 \rangle$	0.210
3	$\langle 0.771, -0.171 \rangle$	0.310
4	$\langle 0.294, 0.147 \rangle$	0.210
5	$\langle 0.170, -0.038 \rangle$	0.310
6	$\langle 0.065, 0.032 \rangle$	

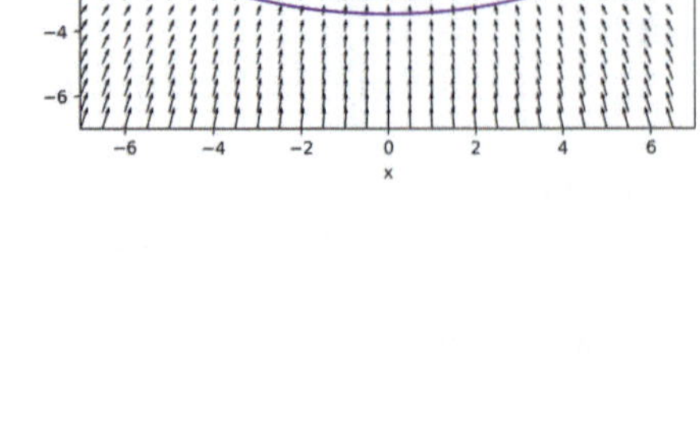

Fig. 9.8 Vector fields and contours for Example 9.17

9.5.2 *The Method of Lagrange Multipliers*

Here we describe the classical method of *Lagrange Multipliers*. To this end, consider the case of the following programming problem.

- Minimize and/or Maximize the objective function

$$f(x_1, \ldots, x_n)$$

- Subject to the constraints

$$g_i(x_1, \ldots, x_n) = b_i \text{ for } i = 1, \ldots, m \text{ with } m < n.$$

In what follows, we assume that f and g_i, for $i = 1, \ldots, m$, have continuous partial derivatives and that the rank of the $m \times n$ Jacobian matrix

$$\left[\frac{\partial g_i}{\partial x_j}\right]_{\substack{i=1,\ldots,m \\ j=1,\ldots,n}}$$

has maximal rank m.

The method of Lagrange Multipliers starts by formulating the *Lagrangian function*

$$h(\mathbf{x}, \lambda) = f(\mathbf{x}) - \sum_{i=1}^{m} \lambda_i (g_i(\mathbf{x}) - b_i)$$

where the new variables $\lambda = (\lambda_1, \ldots, \lambda_m)$ are called *Lagrange multipliers.*

The next step is to solve the system of equations

$$\frac{\partial h}{\partial x_j} = \frac{\partial f}{\partial x_j} - \sum_{i=1}^{m} \lambda_i \frac{\partial g_i}{\partial x_j} = 0 \text{ for } j = 1, \ldots, n$$

and

$$\frac{\partial h}{\partial \lambda_i} = -g_i(\mathbf{x}) + b_i = 0 \text{ for } i = 1, \ldots, m.$$

In doing so, we obtain a collection of solutions $(\mathbf{x}^*, \lambda^*)$ which are such that $\mathbf{x}^*$ is a critical point to the original programming problem. Chapter 10 describes a theory of such *Lagrange multipliers* in a more general context.

Example 9.18. Consider the following nonlinear programming problem.

- Maximize the nonlinear function

$$f(x_1, x_2) = x_1^2 - 2x_2$$

- Subject to the constraint

$$g(x_1, x_2) = x_1^2 + x_2^2 = 1.$$

The *Lagrangian function* is described as

$$h(x_1, x_2) = x_1^2 + 2x_2 - \lambda(x_1^2 + x_2^2 - 1)$$

and thus the equations to be solved are thus

$$\frac{\partial h}{\partial x_1} = 2x_1 - 2\lambda x_1 = 0$$

$$\frac{\partial h}{\partial x_2} = 2 - 2\lambda x_2 = 0$$

and

$$\frac{\partial h}{\partial \lambda} = -(x_1^2 + x_2^2 - 1) = 0.$$

The first equation implies that either $\lambda = 1$ or $x_1 = 0$. If $\lambda = 1$, then the other two equations imply that $x_2 = 1$ and $x_1 = 0$. If $x_1 = 0$, then the third equation implies that $x_2 = \pm 1$. If $x_1 = 0$ then the third equation implies that $x_2 = \pm 1$.

Thus the two critical points for the original problem are $(x_1, x_2) = (0, 1)$ and $(0, -1)$. Respectively, these points are the global maximum and global minimum.

9.6 Karush-Kuhn-Tucker Conditions and Optimal Solutions to Convex Programming Problems

Let $f(x_1, \ldots, x_n)$ be a concave twice differentiable function and suppose that $f_1(x_1, \ldots, x_n), \ldots, f_m(x_1, \ldots, x_n)$ are convex twice differentiable functions. Consider the maximization programming problem

- Maximize the objective function $f(x_1, \ldots, x_n)$
- Subject to the constraints that

$$f_i(x_1, \ldots, x_n) \leq b_i \text{ for } i = 1, \ldots, m$$

and

$$x_i \geq 0 \text{ for } i = 1, \ldots, n.$$

The *Karush-Kuhn-Tucker (KKT) conditions* characterize the optimal solutions to such convex maximization programming problems. The precise result is formulated in the following way.

Theorem 9.19. *Assume that $f(x_1, \ldots, x_n)$ is a concave twice differentiable function and that $f_1(x_1, \ldots, x_n), \ldots, f_m(x_1, \ldots, x_n)$ are convex differentiable functions. Then*

$$\mathbf{x}^* = (x_1^*, \ldots, x_n^*) \in \mathbb{R}^n$$

is an optimal solution to the convex programming problem

- *Maximize the objective function $f(x_1, \ldots, x_n)$*
- *Subject to the constraints that*

$$f_i(x_1, \ldots, x_n) \leq b_i \text{ for } i = 1, \ldots, m$$

and

$$x_i \geq 0 \text{ for } i = 1, \ldots, n$$

if and only if there exist real numbers $v_1^*, \ldots, v_{m+n}^* \geq 0$ which are such that all of the following three KKT conditions *hold true*

$$f_i(\mathbf{x}^*) - b_i \leq 0 \text{ for } i = 1, \ldots, m$$

$$v_i^*(f_i(\mathbf{x}^*) - b_i) = 0 \text{ for } i = 1, \ldots, m$$

$$v_{m+i}^* x_i^* = 0 \text{ for } i = 1, \ldots, n$$

and

$$-\nabla f(\mathbf{x}^*) + v_1^* \nabla f_1(\mathbf{x}^*) + \cdots + v_m^* \nabla f_m(\mathbf{x}^*) - v_{m+1}^* \nabla x_1 - \cdots - v_{m+n} \nabla x_n = \mathbf{0}.$$

The conclusion of Theorem 9.19 follows as a special case of Theorem 10.11. We refer to Section 10.3 for the proof of Theorem 10.11.

Example 9.20. Consider the following two variable maximization programming problem.

- Maximize the objective function

$$f(x_1, x_2) = \ln(x_1 + 1) + x_2$$

- Subject to the constraints

$$2x_1 + x_2 \leq 3$$

and

$$x_1 \geq 0, x_2 \geq 0.$$

Here

$$f(x_1, x_2) = \ln(x_1 + 1) + x_2$$

$$f_1(x_1, x_2) = 2x_1 + x_2$$

$$b_1 = 3$$

$$\nabla f(x_1, x_2) = \left\langle \frac{1}{x_1 + 1}, 1 \right\rangle$$

and

$$\nabla f_1(x_1, x_2) = \langle 2, 1 \rangle .$$

The KKT conditions take the form

$$2x_1^* + x_2^* - 3 \leq 0$$

$$v_1^*(2x_1^* + x_2^* - 3) = 0$$

$$v_2^* x_1^* = 0$$

$$v_3^* x_2^* = 0$$

and

$$-\left\langle \frac{1}{x_1^* + 1}, 1 \right\rangle + v_1^* \langle 2, 1 \rangle - v_2^* \langle 1, 0 \rangle - v_3^* \langle 0, 1 \rangle = \langle 0, 0 \rangle.$$

These above three conditions are satisfied for

$$x_1^* = 0,\ x_2^* = 3,\ v_1^* = 1,\ v_2^* = 1 \text{ and } v_3^* = 0.$$

It follows that

$$\mathbf{x}^* = (0, 3)$$

gives rise to an optimal solution of the given programming problem.

9.6.1 Karush-Kuhn-Tucker Conditions for Quadratic Programming Problems

An important special case of the KKT conditions is their application to *quadratic programming problems*. To begin with, let us recall a general formulation of quadratic programming problems.

- Maximize the quadratic objective function

$$f(x_1, \dots, x_n) = \mathbf{c}\mathbf{x} - \frac{1}{2}\mathbf{x}^T Q\mathbf{x} = \sum_{j=1}^{n} c_j x_j - \frac{1}{2}\sum_{i=1}^{n}\sum_{j=1}^{n} q_{ij} x_i x_j$$

- Subject to the constraints

$$A\mathbf{x} \leq \mathbf{b}$$

and

$$\mathbf{x} \geq \mathbf{0}.$$

Here

$$\mathbf{c} = [c_1, \dots, c_n] \in \mathbb{R}^{1\times n}$$

is an n-dimensional row vector,

$$\mathbf{x} = \begin{bmatrix} x_1 \\ \vdots \\ x_n \end{bmatrix} \in \mathbb{R}^{n\times 1}$$

is an n-dimensional column vector and

$$\mathbf{b} = \begin{bmatrix} b_1 \\ \vdots \\ b_m \end{bmatrix} \in \mathbb{R}^{m\times 1}$$

is an m-dimensional column vector.

On the other hand,

$$Q = [q_{ij}] \in \mathbb{R}^{n\times n}$$

is an $n \times n$ positive semi-definite matrix and

$$A = [a_{ij}] \in \mathbb{R}^{m\times n}$$

is an $m \times n$ matrix. The quadratic objective function $f(x_1, \dots, x_n)$ is a concave function since

$$\mathbf{x}^T Q\mathbf{x} \geq 0 \text{ for all } \mathbf{x}.$$

In other words, $f(x_1, \dots, x_n)$ is a concave function since Q is a positive semi-definite matrix.

In the context just described, namely the case of a quadratic maximization programming problem with concave objective function, it turns out that after introducing suitable slack variables $y_1, \ldots, y_n$ and $v_1, \ldots, v_m$, it turns out that the KKT conditions can be re-expressed as

$$Q\mathbf{x} + A^T\mathbf{u} - \mathbf{y} = \mathbf{c}^T$$

$$A\mathbf{x} + \mathbf{v} = \mathbf{b}$$

$$\mathbf{x} \geq \mathbf{0}, \mathbf{u} \geq \mathbf{0}, \mathbf{y} \geq \mathbf{0}, \mathbf{v} \geq \mathbf{0}$$

and

$$\mathbf{x}^T\mathbf{y} + \mathbf{u}^T\mathbf{v} = 0$$

for

$$\mathbf{u} = \begin{bmatrix} u_1 \\ \vdots \\ u_m \end{bmatrix} \in \mathbb{R}^{m\times 1}.$$

Here,

$$\mathbf{y} = \begin{bmatrix} y_1 \\ \vdots \\ y_n \end{bmatrix} \in \mathbb{R}^{n\times 1} \text{ and } \mathbf{v} = \begin{bmatrix} v_1 \\ \vdots \\ v_m \end{bmatrix} \in \mathbb{R}^{m\times 1}.$$

Note that aside from the final *complementary constraint* all of the other constraints are linear. This is a general feature of KKT conditions for quadratic programming problems that allow for a solution thereof via a modified form of the Simplex Method.

On the other hand, the quadratic programming problems that we have described here fall within the framework of convex programming problems. Hence they can be solved using the general purpose methods for solving convex programming problems. A graphical approach for solving two variable quadratic programming problems is illustrated in Example 11.3.

Example 9.21. Consider the following quadratic maximization programming problem.

- Maximize the objective function

$$f(x_1, x_2) = 15x_1 + 30x_2 + 4x_1x_2 - 2x_1^2 - 4x_2^2$$

- Subject to the constraints

$$x_1 + 2x_2 \leq 30$$

  and

$$x_1, x_2 \geq 0.$$

The matrix

$$Q = \begin{bmatrix} 4 & -4 \\ -4 & 8 \end{bmatrix}$$

is indeed positive semi-definite and so the objective function $f(x_1, x_2)$ is convex.

Then after introducing suitable slack variables y_1, y_2, v_1, the above KKT conditions can be re-expressed in the form

$$Q\mathbf{x} + A^T\mathbf{u} - \mathbf{y} = \mathbf{c}^T$$

$$A\mathbf{x} + \mathbf{v} = \mathbf{b}$$

$$\mathbf{x} \geq 0, \mathbf{u} \geq 0, \mathbf{y} \geq 0, \mathbf{v} \geq 0$$

and

$$\mathbf{x}^T\mathbf{y} + \mathbf{u}^T\mathbf{v} = 0.$$

In more explicit terms the KKT conditions can be re-expressed as

$$4x_1 - 4x_2 + u_1 - y_1 = 15$$

$$-4x_1 + 8x_2 + 2u_1 - y_2 = 30$$

$$x_1 + 2x_2 + v_1 = 30$$

$$x_1 \geq 0, x_2 \geq 0$$

$$u_1 \geq 0, y_1 \geq 0, y_2 \geq 0, v_1 \geq 0$$

and

$$x_1y_1 + v_2y_2 + u_1v_1 = 0.$$

9.7 Portfolio Selection Problem

In this section, we revisit the *Portfolio Selection Problem*, compare with Examples 1.23 and 9.3, and explain how it gives rise to a *quadratic programming problem.*

Especially, recall that investors are concerned with both the *expected return* and the *risk* that is associated with their investments. Suppose that n financial products are being considered for inclusion in the portfolio and let the decision variables x_j, for $j = 1, \ldots, n$, be the number of shares of product j to be included.

Let μ_j and σ_{jj}, respectively, be the estimated *mean* and *variance* of the return on each share of product j where σ_{jj} measures the risk of this product. For $i = 1, \ldots, n$, $i \neq j$, let σ_{ij} be the *covariance* of the return on one share each of product i and product j. In practice, it can be difficult to estimate the covariance values σ_{ij}. However, with suitable hypothesis, values for σ_{ij} can be determined from σ_{ii} and σ_{jj}. A more detailed discussion about this important point is given in Section 9.7.1.

Respectively, the expected value $R(x_1, \ldots, x_n)$ and the variance $V(x_1, \ldots, x_n)$ of the portfolio's total return are given by

$$R(x_1, \ldots, x_n) = \sum_{j=1}^{n} \mu_j x_j$$

and

$$V(x_1, \ldots, x_n) = \sum_{i=1}^{n} \sum_{j=1}^{n} \sigma_{ij} x_i x_j.$$

Note that $V(x_1, \ldots, x_n)$ gives a measure of the *risk* that is associated with the portfolio.

Let L be the minimum expected return. Let P_j be the estimated price for each share of product x_j, for $j = 1, \ldots, n$. Let B be the amount of money that is budgeted for the portfolio. The portfolio selection programming problem that is of interest is then formulated in the following way.

- Minimize the quadratic objective function

$$V(x_1, \ldots, x_n) = \sum_{i=1}^{n} \sum_{j=1}^{n} \sigma_{ij} x_i x_j$$

- Subject to the constraints

$$\sum_{j=1}^{n} \mu_j x_j \geq L$$

$$\sum_{j=1}^{n} P_j x_j \leq B$$

and

$$x_j \geq 0 \text{ for } j = 1, \ldots, n.$$

9.7.1 Construction of Covariance Matrices for the Portfolio Selection Problem

Here, we describe a general approach to the construction of *covariance matrices*. Consider a collection of k random variables $X_1, \ldots, X_k$ and let x_{ij} be the ith independently drawn observation, for $i = 1, \ldots, n$, on the jth random variable X_j, for $j = 1, \ldots, k$. The observations x_{ij} can be arranged into n column vectors

$$\mathbf{x}_i = \begin{bmatrix} x_{i1} \\ \vdots \\ x_{ik} \end{bmatrix} \in \mathbb{R}^k \text{ for } i = 1, \ldots, n.$$

The sample mean vector $\overline{\mathbf{x}}$ has the form

$$\overline{\mathbf{x}} = \begin{bmatrix} \overline{x}_1 \\ \vdots \\ \overline{x}_k \end{bmatrix} \in \mathbb{R}^k$$

where

$$\overline{x}_j = \frac{1}{n}\sum_{i=1}^{n} x_{ij} \text{ for } j = 1, \ldots, k.$$

In other words

$$\overline{\mathbf{x}} = \frac{1}{n}\sum_{i=1}^{n} \mathbf{x}_i = \begin{bmatrix} \overline{x}_1 \\ \vdots \\ \overline{x}_k \end{bmatrix}.$$

In this context, the *sample covariance matrix* can be defined as the $k \times k$ matrix

$$Q = \left[q_{j\ell}\right]_{j,\ell=1,\ldots,k}$$

where

$$Q = \frac{1}{n-1}\sum_{i=1}^{n}(\mathbf{x}_i - \overline{\mathbf{x}})(\overline{\mathbf{x}}_i - \overline{\mathbf{x}})^T;$$

that is

$$q_{j\ell} = \frac{1}{n-1}\sum_{i=1}^{n}(x_{ij} - \overline{x}_j)(x_{i\ell} - \overline{x}_\ell).$$

In essence, the construction of a suitable covariance matrix Q amounts to appropriate choices of observations x_{ij}. The variance $V(x_1, \ldots, x_n)$ will be a convex function as the covariance matrix Q is positive semi-definite.

9.8 Separable Programming

Separable programming is a special case of *convex programming*, where the objective function $f(x_1, \ldots, x_n)$ and the constraint functions $g_i(x_1, \ldots, x_n)$ are all separable functions. For our purposes, by a *separable function*, is meant a function in which each term involves just a single variable.

So, for example, if an objective function $f(x_1, \ldots, x_n)$ is separable, then it can be expressed as

$$f(x_1, \ldots, x_n) = \sum_{j=1}^{n} f_j(x_j)$$

where each $f_j(x_j)$ includes terms that only involve x_j.

In what follows, we describe a special case of separable programming problems. They can be solved by reduction to the Simplex Method. The approach also indicates a method to reformulate certain programming problems with piecewise linear separable objective functions as suitable linear programming problems.

For our purposes, we assume that each of the constraint functions $g_i(x_1, \ldots, x_n)$ are linear and that the objective function $f(x_1, \ldots, x_n)$ is of the form

$$f(x_1, \ldots, x_n) = \sum_{j=1}^{n} f_j(x_j)$$

where each $f_j(x_j)$ is concave or piecewise linear concave.

The idea, is to approximate a given separable linear programming problem, with objective function described as above, by a suitable linear programming model. To this end, let n_j be the number of line segments in the piecewise linear function that approximates the given concave function $f_j(x_j)$.

Then

$$x_j = \sum_{k=1}^{n_j} x_{jk}$$

and

$$f_j(x_j) = \sum_{k=1}^{n_j} s_{jk} x_{jk}.$$

The resulting linear programming model that is of interest takes the following form.

- Maximize the linear objective function

$$Z = \sum_{j=1}^{n} \left(\sum_{k=1}^{n_j} s_{jk} x_{jk} \right)$$

- Subject to the constraints

$$\sum_{j=1}^{n} a_{ij} \left(\sum_{k=1}^{n_j} x_{jk} \right) \le b_j$$

for $i = 1, \ldots, m$,

$$x_{jk} \le u_{jk}$$

for $k = 1, \ldots, n_j$ and $j = 1, \ldots, n$, and

$$x_{jk} \ge 0$$

for $k = 1, \ldots, n_j$ and $j = 1, \ldots, n$.

We illustrate the above discussion in Example 9.22 below.

Example 9.22. Suppose that two products, Product 1 and Product 2, have respective production rates x_1 and x_2 which are subject to the production constraints

$$\begin{aligned} x_1 \qquad &\le 4 \\ 2x_2 &\le 12 \\ 3x_1 + 2x_2 &\le 18 \end{aligned}$$

and

$$x_1, x_2 \ge 0.$$

Suppose that the respective rates of profit as a function of production rate is described as in the following two graphs. (See Figure 9.9.)

The production rate for Product 1 can be thought of as

$$x_1 = x_{11} + x_{12}$$

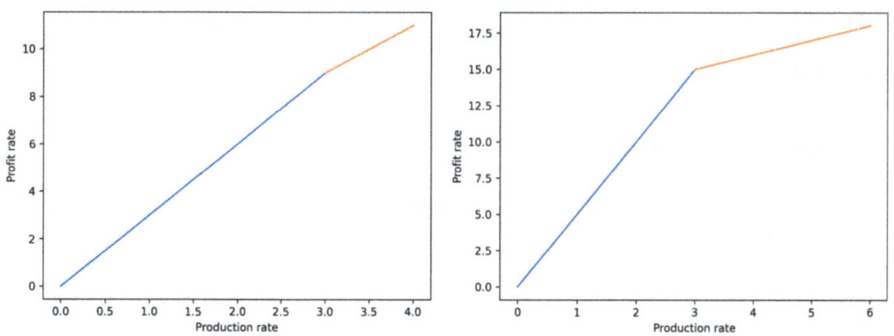

Fig. 9.9 Profit rates for Example 9.22

where x_{11} is the production rate on regular time and where x_{12} is the production rate from usage over time.

Similarly, the production rate for Product 2 can be thought of as

$$x_2 = x_{21} + x_{22}$$

where x_{21} is the production rate on regular time and where x_{22} is the production rate from usage over time.

Altogether, a linear programming model to maximize profit subject to the given production constraints is formulated in the following way.

- Maximize the linear objective function

$$Z = 3x_{11} + 2x_{12} + 5x_{21} + x_{22}$$

- Subject to the constraints

$$x_{11} + x_{21} \leq 4$$

$$2(x_{21} + x_{22}) \leq 12$$

$$3(x_{11} + x_{12}) + 2(x_{21} + x_{22}) \leq 18$$

$$x_{11} \leq 3, x_{12} \leq 1, x_{21} \leq 3, x_{22} \leq 3$$

and

$$x_{11} \geq 0, x_{12} \geq 0, x_{21} \geq 0, x_{22} \geq 0.$$

Solving this linear programming problem, see Example 11.4, the maximum value of the objective function is

$$Z = 23.$$

This is achieved when $x_{11} = 1, x_{12} = 1, x_{21} = 3$ and $x_{22} = 3$.

Exercises

9.1 Apply Newton's Method with

$$\epsilon = 0.001$$

and

$$x_1 = 1$$

to the unconstrained maximization problem

- Maximize the objective function

$$f(x) = 48x^5 + 42x^3 + 3.5x - 16x^6 - 61x^4 - 16.5x^2.$$

9.2 Consider the following nonlinear programming problem

- Maximize the objective function

$$f(x_1, x_2) = \ln(x_1 + 1) - x_2^2$$

- Subject to the constraints

$$x_1 + 2x_2 \leq 3$$

and

$$x_1, x_2 \geq 0.$$

(i) Verify that this problem is a convex programming problem.
(ii) Use the KKT conditions to derive an optimal solution.

9.3 Solve the quadratic programming problem

- Maximize the objective function

$$f(x_1, x_2) = 2x_1 + 3x_2 - x_1^2 - x_2^2$$

- Subject to the constraints

$$x_1 + x_2 \leq 2$$

and

$$x_1, x_2 \geq 0.$$

9.4 Suppose that Company A has two new products that will compete with two new products from Company B. Suppose that the linear programming model

- Maximize the objective function

$$f(x_1, x_2) = 4x_1 + 6x_2$$

- Subject to the constraints

$$\begin{aligned} x_1 + 3x_2 &\leq 8 \\ 5x_1 + 2x_2 &\leq 14 \end{aligned}$$

and

$$x_1, x_2 \geq 0$$

has been formulated to determine the most profitable product mix. Suppose that the units for the objective function are given in hundreds of units of dollars.

However, because of competition from Company B, suppose that management now realizes that Company A will need to make a strong marketing effort to generation substantial sales of these products. Indeed, suppose that it is estimated that achieving a production and sales rate of x_1 units of Product 1 per week will require weekly marketing costs of x_1^3 hundred dollars; the corresponding marketing costs for Product 2 are estimated to be $2x_2^2$ hundred dollars.

(i) Write down the objective function for this revised model.
(ii) Determine the KKT conditions for this revised model.
(iii) Verify that $(x_1, x_2) = (\frac{2}{\sqrt{3}}, \frac{3}{2})$ is an optimal solution using the KKT conditions.
(iv) Use a suitable separable programming model based on production rates of $0, 1, 2, 3$ to formulate an approximate linear programming model for this problem.

9.5 Suppose that an investor wishes to invest in six potential stock choices $x_1, x_2, x_3, x_4, x_5, x_6$. By studying historical stock market data, suppose that the investor has constructed the following 6×6 covariance matrix

$$\text{Covariance Matrix} = \begin{bmatrix} 0.032 & 0.005 & 0.03 & -0.031 & -0.027 & 0.01 \\ 0.005 & 0.1 & 0.085 & -0.07 & -0.05 & 0.02 \\ 0.03 & 0.085 & 0.333 & -0.11 & -0.02 & 0.042 \\ -0.031 & -0.07 & -0.11 & 0.125 & 0.05 & -0.06 \\ -0.027 & -0.05 & -0.02 & 0.05 & 0.065 & -0.02 \\ 0.01 & 0.02 & 0.042 & -0.06 & -0.02 & 0.08 \end{bmatrix}.$$

Here, the i, j entry in the above matrix is the i, jth covariance $\sigma_{i,j}$.

Now, based on current market trends, suppose that stock 1 is expected to rise from \$60 per share to \$72 per share, over the next year, whereas stock 2 is expected to rise from \$127 per share to \$275. Further, stock 3 is expected to rise from \$4 per share to \$8, stock 4 is expected to rise from \$50 per share to \$75 per share and stock 5 is expected to rise from \$90 per share to \$105. Finally, stock 6 is expected to rise, over the next year, from \$20 per share to \$26.

Assume that the investor has \$200, 000 at the start of the year and that the investor wishes to have at least \$225, 000 in the investment portfolio at the end of the year.

(i) Give an explicit mathematical description of the portfolio's expected return

$$R(x_1, \ldots, x_n) = \sum_{j=1}^{n} \mu_j x_j.$$

(ii) Give an explicit mathematical description of the portfolio's expected variance

$$V(x_1, \ldots, x_n) = \sum_{i=1}^{n} \sum_{j=1}^{n} \sigma_{i,j} x_i x_j.$$

(iii) Formulate a quadratic minimization programming problem to achieve a minimum risk investment strategy.

Chapter 10
More General Convex Functions, Lagrangians and KKT Conditions

Overview

- More general convex functions.
- Differential ordinary programming problems.
- Lagrangians and Karush-Kuhn-Tucker conditions.

10.1 Preliminaries for Convex Functions

In what follows, we expose the most basic theory that surrounds *convex programming problems*. Our approach is mostly based on that of [14]. The aim is to develop the theoretical foundations for *differential ordinary convex programming problems* and their associated Lagrangians and Karush-Kuhn-Tucker conditions. In this regard, the main result that we prove is Theorem 10.11 (see Section 10.3). Theorem 10.11 gives the main Duality Theorem for convex programming problems.

Here, and in Section 10.2, we consider a more general context for working with convex functions. This is required in order to state and prove Theorem 10.11.

For this purpose, suppose that

$$S \subseteq \mathbb{R}^n$$

is a nonempty subset and consider a function

$$f : S \to \mathbb{R} \bigcup \{\pm\infty\}.$$

Define its *epigraph* to be

$$\operatorname{epi}(f) := \{(\mathbf{x}, a) : \mathbf{x} \in S, a \in \mathbb{R} \text{ and } f(\mathbf{x}) \leq a\} \subseteq \mathbb{R}^{n+1}.$$

N. Grieve, *An Introduction to Mathematical Programming and Network Science*, Springer Undergraduate Texts in Mathematics and Technology,
https://doi.org/10.1007/978-3-032-13330-4_10

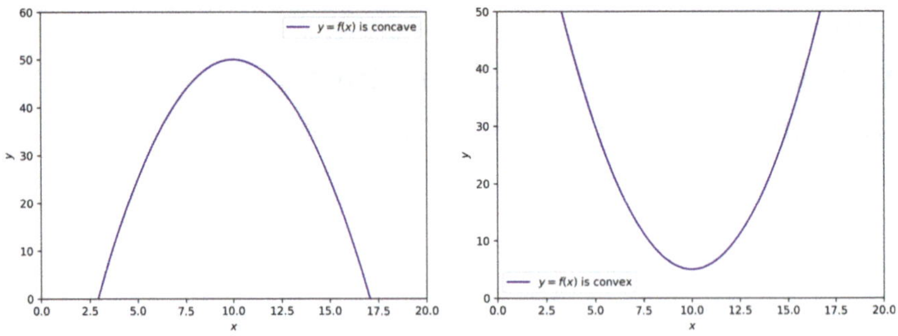

Fig. 10.1 Convex and concave functions for Example 10.1

We say that f is *convex* if epi(f) is a convex subset of $\mathbb{R}^{n+1}$; f is *concave* if $-f$ is convex.

Example 10.1. The intuition behind these more general concepts of convex and concave functions arise from the classical case of graphs of single variable smooth real valued functions. This is illustrated in Figure 10.1.

The *effective domain* of a convex function

$$f : S \to \mathbb{R} \bigcup \{\pm\infty\}$$

is defined to be

$$\operatorname{dom}(f) := \{\mathbf{x} \in S : f(\mathbf{x}) < +\infty\} \subseteq \mathbb{R}^n;$$

this is a convex subset of $\mathbb{R}^n$. Indeed, for example, it is the projection of epi(f) onto $\mathbb{R}^n$.

An *affine function* on S is a function

$$f : S \to \mathbb{R} \bigcup \{\pm\infty\}$$

which takes on finite values and which is both convex and concave.

A convex function

$$f : S \to \mathbb{R} \bigcup \{\pm\infty\}$$

is *proper* if

$$f(\mathbf{x}) < +\infty$$

for at least one $\mathbf{x} \in S$ and

$$f(\mathbf{x}) > -\infty$$

for all $\mathbf{x} \in S$.

Example 10.2. Recall that affine linear functions

$$f: \mathbb{R}^n \to \mathbb{R}$$

are defined by the condition that

$$f(\mathbf{x}) = a_1 x_1 + \cdots + a_n x_n - d$$

for $a_1, \ldots, a_n \in \mathbb{R}$ and $d \in \mathbb{R}$. These are proper affine functions in the sense that is described above.

To place matters into perspective, we state and prove some of the most basic results that surround the theory of convex functions. The key point, is that convex functions exhibit an important interpolation property. This is expressed in Lemma 10.3 below.

Lemma 10.3. *Suppose that*

$$S \subseteq \mathbb{R}^n$$

is a nonempty subset and consider a function

$$f: S \to \mathbb{R} \bigcup \{\pm\infty\}.$$

Then f is convex on S if and only if

$$f((1-\lambda)\mathbf{x} + \lambda\mathbf{y}) \leq (1-\lambda)a + \lambda b$$

whenever $\mathbf{x}, \mathbf{y} \in S$, $f(\mathbf{x}) \leq a \in \mathbb{R}$, $f(\mathbf{y}) \leq b \in \mathbb{R}$ and $0 \leq \lambda \leq 1$.

Proof. By definition, f is convex on S if and only if

$$(1-\lambda)(\mathbf{x}, a) + \lambda(\mathbf{y}, b) \in \text{epi}(f)$$

whenever

$$(\mathbf{x}, a), (\mathbf{y}, b) \in \text{epi}(f) \text{ and } 0 \leq \lambda \leq 1.$$

Equivalently, this means exactly that

$$f((1-\lambda)\mathbf{x} + \lambda\mathbf{y}) \leq (1-\lambda)a + \lambda b$$

whenever $\mathbf{x}, \mathbf{y} \in S$, $f(\mathbf{x}) \leq a \in \mathbb{R}$, $f(\mathbf{y}) \leq b \in \mathbb{R}$ and $0 \leq \lambda \leq 1$. □

Two immediate consequences of Lemma 10.3 are Theorems 10.4 and 10.5 below.

Theorem 10.4. *Suppose that*

$$C \subseteq \mathbb{R}^n$$

is a nonempty convex subset. Then f is convex on C if and only if

$$f((1-\lambda)\mathbf{x}+\lambda\mathbf{y}) \leq (1-\lambda)f(\mathbf{x})+\lambda f(\mathbf{y}) \text{ for } 0<\lambda<1 \text{ for all } \mathbf{x},\mathbf{y} \in C.$$

Proof. This follows immediately from Lemma 10.3, upon setting $S = C$. □

Theorem 10.5. *Consider a function*

$$f\colon \mathbb{R}^n \to [-\infty, +\infty].$$

Then f is convex if and only if

$$f((1-\lambda)\mathbf{x}+\lambda\mathbf{y}) < (1-\lambda)a+\lambda b \text{ for } 0<\lambda<1$$

whenever $f(\mathbf{x}) < a$ and $f(\mathbf{y}) < b$.

Proof. This follows immediately from Lemma 10.3, upon setting $S = \mathbb{R}$. □

In what follows, we let $\langle\cdot,\cdot\rangle$ denote the standard inner product on $\mathbb{R}^n$. Consider a function

$$f\colon \mathbb{R}^n \to \mathbb{R}\bigcup\{\pm\infty\}$$

and suppose that $\mathbf{x} \in \mathbb{R}$ is such that $f(\mathbf{x})$ is finite.

Then, f is said to be *differentiable* at $\mathbf{x}$ if and only if there exists a vector $\mathbf{x}^* \in \mathbb{R}^n$ which is such that

$$\lim_{\mathbf{z}\to\mathbf{x}} \frac{f(\mathbf{z})-f(\mathbf{x})-\langle \mathbf{x}^*, \mathbf{z}-\mathbf{x}\rangle}{||\mathbf{z}-\mathbf{x}||} = 0.$$

Such an $\mathbf{x}^*$, if it exists, is called the *gradient* of f at $\mathbf{x}$ and is denoted by $\nabla f(\mathbf{x})$. Among other things, denoting by $\mathbf{e}_1, \ldots, \mathbf{e}_n \in \mathbb{R}^n$ the standard basis vectors for $\mathbb{R}^n$, it follows that

$$\left\langle \nabla f(\mathbf{x}), \mathbf{e}_j \right\rangle = \frac{\partial f}{\partial x_j}(\mathbf{x})$$

for $j = 1, \ldots, n$. Further, it follows that

$$\nabla f(\mathbf{x}) = \left\langle \frac{\partial f}{\partial x_1}(\mathbf{x}), \ldots, \frac{\partial f}{\partial x_n}(\mathbf{x}) \right\rangle.$$

By this notation, we may denote by $f'(\mathbf{x};\mathbf{y})$ the linear functional

$$f'(\mathbf{x};\mathbf{y}) = \langle \nabla f(\mathbf{x}), \mathbf{y}\rangle.$$

This concept of differentiability, as we have defined here, extends that which is familiar from multivariable calculus.

The most significant classical examples of convex functions arise as consequences of the following theorem.

Theorem 10.6. *Consider a twice differentiable function*

$$f : (a, b) \to \mathbb{R},$$

for $(a, b) \subseteq \mathbb{R}$ *an open interval. Then* f *is convex if and only if its second derivative,* f''*, is nonnegative throughout* (a, b)*.*

Proof. $\Leftarrow$]. Suppose first that $f''(x) \geq 0$ for $x \in (a, b)$. Then f' is not decreasing on (a, b). Thus for all $a < y < b$, $0 < \lambda < 1$ and setting

$$z = (1 - \lambda)x + \lambda y$$

it then holds true that

$$f(z) - f(x) = \int_x^z f'(t)\mathrm{d}t \leq f'(z)(z - x)$$

and

$$f(y) - f(z) = \int_z^y f'(t)\mathrm{d}t \geq f'(z)(y - z).$$

Then, since

$$z - x = \lambda(y - x) \text{ and } y - z = (1 - \lambda)(y - x)$$

it follows that

$$f(z) \leq f(x) + \lambda f'(z)(y - x)$$

and

$$f(z) \leq f(y) - (1 - \lambda) f'(z)(y - x).$$

Upon multiplying these two above inequalities, respectively, by $(1 - \lambda)$ and λ, and then adding the results together the conclusion is that

$$(1 - \lambda) f(z) + \lambda f(z) \leq (1 - \lambda) f(x) + \lambda f(y)$$

whence the convexity of f on (a, b) is readily deduced.

$\Rightarrow$]. Conversely, suppose by way of contradiction that f'' is not nonnegative on (a, b). Then, by continuity, it follows that f'' must be negative on a certain subinterval on (a, b). But then on this subinterval if $z = (1 - \lambda)x + \lambda y$, for $0 < \lambda < 1$, then

$$f(z) - f(x) > f'(z)(z - x)$$

and

$$f(y) - f(z) < f'(z)(y - z)$$

whence it follows that

$$f((1 - \lambda)x + \lambda y) > (1 - \lambda)f(x) + \lambda f(y),$$

for some $0 < \lambda < 1$. The conclusion is that f is not convex on (a, b). This is a contradiction. □

Theorem 10.6 implies its multidimensional generalization.

Theorem 10.7. *Consider a twice differentiable function*

$$f : C \to \mathbb{R}^n$$

on a nonempty open convex set $C \subseteq \mathbb{R}^n$. Then f is convex on C if and only if its Hessian matrix

$$Q_{\mathbf{x}} := \left[q_{ij}(\mathbf{x})\right] \textit{ for } q_{ij}(\mathbf{x}) := \left.\frac{\partial^2 f}{\partial x_i \partial x_j}\right|_{\mathbf{x}=(x_1,\ldots,x_n)}$$

is positive semi-definite for all $\mathbf{x} \in C$.

Proof. The convexity of f on C is equivalent to the convexity of the restriction of f to each line segment in C. In turn, this is equivalent to the convexity of

$$g(\lambda) := f(\mathbf{y} + \lambda \mathbf{z})$$

on the set

$$\{\lambda : \mathbf{y} + \lambda \mathbf{z} \in C \text{ for all } \mathbf{y} \in C \text{ and } \mathbf{z} \in \mathbb{R}^n\}.$$

Thus, by Theorem 10.6, it follows that g is convex for each $y \in C$ and $z \in \mathbb{R}^n$ if and only if

$$\langle z, Q_{\mathbf{x}} z\rangle \geq 0$$

for all $\mathbf{z} \in \mathbb{R}^n$ and $\mathbf{x} \in C$. □

For later use, we define the concept of *affine hull* of a subset

$$S \subseteq \mathbb{R}^n.$$

It is denoted by $\text{aff}(S)$ and consists of all vectors of the form $\lambda_1\mathbf{x}_1 + \cdots + \lambda_m\mathbf{x}_m$ where $\mathbf{x}_i \in S$ and where $\lambda_1, \ldots, \lambda_m \in \mathbb{R}$ are such that $\lambda_1 + \cdots + \lambda_m = 1$.

Further, we define the *relative interior* of a convex subset

$$C \subseteq \mathbb{R}^n.$$

This is denoted by $\text{r.i.}(C)$ and is defined to be the set of all points $\mathbf{x} \in \text{aff}(C)$ which have the property that there exists $\epsilon > 0$ which is such that $\mathbf{y} \in C$ if $\mathbf{y} \in \text{aff}(C)$ and $\mathbf{y}$ has distance $d(\mathbf{x}, \mathbf{y})$ at most ϵ from $\mathbf{x}$.

10.2 Preliminaries about Subgradients

In this section, we discuss the *subgradients* of convex functions. The main results are Theorems 10.8 and 10.9 which we use in Section 10.3 to prove Theorem 10.11.

Let $f(\mathbf{x})$ be a closed proper convex function on $\mathbb{R}^n$. The *subdifferential mapping* $\partial f(\mathbf{x})$, for $\mathbf{x} \in \mathbb{R}^n$, assigns to $f(\mathbf{x})$ a certain closed convex subset

$$\partial f(\mathbf{x}) \subseteq \mathbb{R}^n.$$

More precisely, a vector $\mathbf{x}^* \in \mathbb{R}^n$ is a *subgradient* of a convex function $f(\mathbf{x})$ at a point $\mathbf{x} \in \mathbb{R}^n$ if

$$f(\mathbf{z}) \geq f(\mathbf{x}) + \langle \mathbf{x}^*, \mathbf{z} - \mathbf{x} \rangle \text{ for all } \mathbf{z}.$$

Define

$$\partial f(\mathbf{x}) := \left\{ \mathbf{x}^* \in \mathbb{R}^n : \mathbf{x}^* \text{ is a } \mathit{subgradient} \text{ of } f(\mathbf{x}) \text{ at } \mathbf{x} \right\}.$$

Then, $\mathbf{x}^* \in \partial f(\mathbf{x})$ if and only if $\langle \mathbf{z}, \mathbf{x}^* \rangle - f(\mathbf{z})$ achieves its supremum in $\mathbf{z}$ at $\mathbf{z} = \mathbf{x}$.

Theorem 10.8. *Let $f_1(\mathbf{x}), \ldots, f_m(\mathbf{x})$ be proper convex functions on $\mathbb{R}^n$ and let*

$$f = f_1(\mathbf{x}) + \cdots + f_m(\mathbf{x}).$$

Then

$$\partial f(\mathbf{x}) \supseteq \partial f_1(\mathbf{x}) + \cdots + \partial f_m(\mathbf{x}) \textit{ for all } \mathbf{x}.$$

Further, if

$$\varnothing \neq \bigcap_{i=1}^{m} \text{r.i.}(\text{dom } f_i(\mathbf{x})) \textit{ for } i = 1, \ldots, m$$

then

$$\partial f(\mathbf{x}) = \partial f_1(\mathbf{x}) + \cdots + \partial f_m(\mathbf{x}) \textit{ for all } \mathbf{x}.$$

Proof. If $\mathbf{x}^* = \mathbf{x}_1^* + \cdots + \mathbf{x}_m^*$, where $\mathbf{x}_i^* \in \partial f_i(\mathbf{x})$, then we have for all $\mathbf{z}$ that

$$\begin{aligned} f(\mathbf{z}) = f_1(\mathbf{z}) + \cdots + f_m(\mathbf{z}) &\geq f_1(\mathbf{x}) + \langle \mathbf{z} - \mathbf{x}, \mathbf{x}_1^* \rangle + \cdots + f_m(\mathbf{x}) + \langle \mathbf{z} - \mathbf{x}, \mathbf{x}_m^* \rangle \\ &= f(\mathbf{x}) + \langle \mathbf{z} - \mathbf{x}, \mathbf{x}_1^* + \cdots + \mathbf{x}_m^* \rangle = f(\mathbf{x}) + \langle \mathbf{z} - \mathbf{x}, \mathbf{x}^* \rangle. \end{aligned}$$

Thus $\mathbf{x}^* \in \partial f(\mathbf{x})$.

If

$$\varnothing \neq \bigcap_{i=1}^{m} \mathrm{r.\,i.}(\mathrm{dom}(f_i))$$

then

$$\mathbf{x}^* \in \partial f(x)$$

if and only if

$$\langle \mathbf{x}, \mathbf{x}^* \rangle = f_1(\mathbf{x}) + \cdots + f_m(\mathbf{x}) + \inf_{\mathbf{x}^* = \mathbf{x}_1^* + \cdots + \mathbf{x}_m^*} \{ f_1^*(\mathbf{x}_1^*) + \cdots + f_m^*(\mathbf{x}_m^*) \}.$$

where

$$f_i^*(\mathbf{x}_i^*) = \sup_{\mathbf{z}} \big(\langle \mathbf{z}, \mathbf{x}_i^* \rangle - f(\mathbf{z}) \big)$$

for $i = 1, \ldots, m$ and where for each $\mathbf{x}^*$ the infimum is attained by some $\mathbf{x}_1^*, \ldots, \mathbf{x}_m^*$.

It follows that $\partial f(\mathbf{x})$ consists of the vectors of the form $\mathbf{x}_1^* + \cdots + \mathbf{x}_m^*$ which are such that

$$\langle \mathbf{x}, \mathbf{x}_1^* \rangle + \cdots + \langle \mathbf{x}, \mathbf{x}_m^* \rangle = f_1(\mathbf{x}) + \cdots + f_m(\mathbf{x}) + f_1^*(\mathbf{x}_1^*) + \cdots + f_m^*(\mathbf{x}_m^*).$$

But there is always the inequality

$$\langle \mathbf{x}, \mathbf{x}_i^* \rangle \leq f_i(\mathbf{x}) + f_i^*(\mathbf{x}_i^*)$$

with equality if and only if

$$\mathbf{x}_i^* \in \partial f_i(\mathbf{x}) \text{ for } i = 1, \ldots, m.$$

Thus, $\partial f(\mathbf{x})$ is the same as

$$\partial f_1(\mathbf{x}) + \cdots + \partial f_m(\mathbf{x}).$$

□

Theorem 10.9. *Let f be a convex function and let $\mathbf{x}$ be a point where f is finite. Then the following assertions hold true.*

(i) A point $\mathbf{x}^*$ is a subgradient of f at $\mathbf{x}$ if and only if

$$f'(\mathbf{x};\mathbf{y}) \geq \langle \mathbf{x}^*, \mathbf{y} \rangle \text{ for all } \mathbf{y}.$$

(ii) If $f(\mathbf{x})$ is differentiable at $\mathbf{x}$, then $\nabla f(\mathbf{x})$ is the unique subgradient of f at $\mathbf{x}$ so that

$$f(\mathbf{z}) \geq f(\mathbf{x}) + \langle \nabla f(\mathbf{x}), \mathbf{z} - \mathbf{x} \rangle \text{ for all } \mathbf{z}.$$

Proof. First of all, recall that

$$f'(\mathbf{x};\mathbf{y}) = \langle \nabla f(\mathbf{x}), \mathbf{y} \rangle.$$

Then, for item (i), setting

$$\mathbf{z} = \mathbf{x} + \lambda \mathbf{y},$$

the subgradient inequality

$$f(\mathbf{z}) \geq f(\mathbf{x}) + \langle \mathbf{x}^*, \mathbf{z} - \mathbf{x} \rangle$$

can be re-expressed as

$$\frac{f(\mathbf{x} + \lambda \mathbf{y}) - f(\mathbf{x})}{\lambda} \geq \langle \mathbf{x}^*, \mathbf{y} \rangle$$

for all $\mathbf{y}$ and all $\lambda > 0$. Since the difference quotient decreases to $f'(\mathbf{x};\mathbf{y})$ as $\lambda \to 0$, the desired inequality follows.

On the other hand, for item (ii), suppose that f is differentiable at $\mathbf{x}$. Then $f'(\mathbf{x};\cdot)$ is the linear function $\langle \nabla f(\mathbf{x}), \cdot \rangle$. By Theorem 10.9 (i), the subgradients of $\mathbf{x}$ are those vectors $\mathbf{x}^*$ which are such that

$$\langle \nabla f(\mathbf{x}), \mathbf{y} \rangle \geq \langle \mathbf{x}^*, \mathbf{y} \rangle \text{ for all } \mathbf{y}.$$

This condition is satisfied if and only if

$$\mathbf{x}^* = \nabla f(\mathbf{x}).$$

Thus, $\nabla f(\mathbf{x})$ is the unique subgradient of f at $\mathbf{x}$. □

10.3 Ordinary Convex Programming Problems

The aim of this section is to state and prove Theorem 10.11. This is the *Duality Theorem* for convex programming problems.

For our purposes here, an *ordinary convex programming problem*, is a minimization programming problem of the following form.

- Minimize the convex objective function

$$f_0 = f : C \to \mathbb{R} \bigcup \{\pm\infty\}$$

  for $C \subseteq \mathbb{R}^n$ a nonempty convex subset
- Subject to the constraints

$$f_1(\mathbf{x}) \leq 0, \ldots, f_r(\mathbf{x}) \leq 0, f_{r+1}(\mathbf{x}) = 0, \ldots, f_{r+m}(\mathbf{x}) = 0.$$

  Here, $f_1(\mathbf{x}), \ldots, f_r(\mathbf{x})$, respectively, $f_{r+1}(\mathbf{x}), \ldots, f_{r+m}(\mathbf{x})$ are convex, respectively, affine, functions on C and taking values in $\mathbb{R} \bigcup \{\pm\infty\}$.

An ordinary convex programming problem is *differentiable* if f and $f_i(\mathbf{x})$ are differentiable for all $i = 1, \ldots, r + m$.

Remark 10.10. For a given ordinary convex programming problem if $r = m$, then there are no equality constraints. On the other hand, if $r = 0$, then there are no inequality constraints.

When considering ordinary convex programming problems, for technical convenience, we assume without loss of generality, in what follows, that

- $f(\mathbf{x})$ is a proper convex function on all of $\mathbb{R}^n$ with the property that

$$\mathrm{dom}(f(\mathbf{x})) = C;$$

- $f_1(\mathbf{x}), \ldots, f_r(\mathbf{x})$ are proper convex functions on all of $\mathbb{R}^n$, with $\mathrm{r.\,i.}(\mathrm{dom}(f_i))$ and $\mathrm{r.\,i.}(C)$ the respective relative interiors of $\mathrm{dom}(f_i)$, for $i = 1, \ldots, r$, and C, which are such that

$$\mathrm{r.\,i.}(\mathrm{dom}(f_i)) \supset \mathrm{r.\,i.}(C), \text{ and } \mathrm{dom}(f_i)) \supset C; \text{ and}$$

-

$$f_{r+1}(\mathbf{x}), \ldots, f_{r+m}(\mathbf{x}) \text{ are affine on all of } \mathbb{R}^n.$$

The *Lagrangian* of an ordinary convex programming problem is defined to be

$$L(\mathbf{v}, \mathbf{x}) := f(\mathbf{x}) + v_1 f_1(\mathbf{x}) + \cdots + v_{r+m} f_{r+m}(\mathbf{x})$$

where

$$\mathbf{v} = (v_1, \ldots, v_{r+m}) \in \mathbb{R}^{r+m}_{\geq 0} \text{ and } \mathbf{x} = (x_1, \ldots, x_n) \in C.$$

We set

$$L(\mathbf{v}, \mathbf{x}) = -\infty \text{ if } \mathbf{v} = (v_1, \ldots, v_{r+m}) \notin \mathbb{R}^{r+m}_{\geq 0}$$

and

$$L(\mathbf{v}, \mathbf{x}) = \infty \text{ if } \mathbf{x} = (x_1, \ldots, x_n) \notin C.$$

The nonnegative real number

$$v_i \in \mathbb{R}_{\geq 0} \text{ for } i = 1, \ldots, r + m$$

is the ith *Lagrange multiplier* that is associated with the given programming problem's ith constraint. The Lagrangian $L(\mathbf{v}, \mathbf{x})$ is concave in $\mathbf{v}$, for each $\mathbf{x}$, and convex in $\mathbf{x}$, for each $\mathbf{v}$.

A vector pair

$$(\mathbf{v}^*, \mathbf{x}^*) \in \mathbb{R}^{r+m} \times \mathbb{R}^n$$

is a *saddle point* of $L(\mathbf{v}, \mathbf{x})$ with respect to maximizing $\mathbf{v}$ and minimizing $\mathbf{x}$ if

$$L(\mathbf{v}, \mathbf{x}^*) \leq L(\mathbf{v}^*, \mathbf{x}^*) \leq L(\mathbf{v}^*, \mathbf{x}) \text{ for all } \mathbf{v} \in \mathbb{R}^{r+m} \text{ and all } \mathbf{x} \in \mathbb{R}^n.$$

We want to prove the *Duality Theorem* for ordinary differential convex programming problems, see Theorem 10.11, which characterizes their optimal solutions in terms of the saddle points of the corresponding Lagrangian and KKT equations.

First we say that

$$\mathbf{v} = (v_1, \ldots, v_{r+m}) \in \mathbb{R}^{r+m}$$

is a *vector of KKT coefficients* for a given ordinary convex programming problem if

- $v_i \geq 0$ for $i = 1, \ldots, r + m$; and
- $\inf(h(\mathbf{x})) = \inf(f(\mathbf{x})) < \infty$, for $h(\mathbf{x}) := f(\mathbf{x}) + v_1 f_1(\mathbf{x}) + \cdots + v_{r+m} f_{r+m}(\mathbf{x})$.

With these preliminaries at hand, the *Duality Theorem* for convex programming problems is formulated in the following way.

Theorem 10.11 (Duality Theorem for convex programming problems). *Consider the case of a differential ordinary convex programming problem. Let* $\mathbf{v}^* \in \mathbb{R}^{r+m}$ *and* $\mathbf{x}^* \in \mathbb{R}^n$*. Then* $\mathbf{v}^*$ *is a KKT vector and* $\mathbf{x}^*$ *is an optimal solution if and only if*

$$(\mathbf{v}^*, \mathbf{x}^*) \in \mathbb{R}^{r+m} \times \mathbb{R}^n$$

is a saddle point of the Lagrangian $L(\mathbf{v}, \mathbf{x})$. *Moreover, this condition holds if and only if* $\mathbf{x}^*$ *and the components* v_i^* *of* $\mathbf{v}^*$ *satisfy the three conditions that*

$$v_i^* \geq 0,\ f_i(\mathbf{x}^*) \leq 0 \text{ and } v_i^* f_i(\mathbf{x}^*) = 0 \text{ for } i = 1, \ldots, r. \tag{10.1}$$

$$f_i(\mathbf{x}^*) = 0 \text{ for } i = r+1, \ldots, r+m. \tag{10.2}$$

$$\nabla f(\mathbf{x}^*) + v_1^* \nabla f_1(\mathbf{x}^*) + \cdots + v_{r+m}^* \nabla f_{r+m}(\mathbf{x}^*) = \mathbf{0}. \tag{10.3}$$

Proof. By definition

$$(\mathbf{v}^*, \mathbf{x}^*) \in \mathbb{R}^{r+m} \times \mathbb{R}^n$$

is a saddle point of L if and only if

$$\sup_{\mathbf{v}} L(\mathbf{v}, \mathbf{x}^*) = L(\mathbf{v}^*, \mathbf{x}^*) = \inf_{\mathbf{x}} (\mathbf{v}^*, \mathbf{x}).$$

However, the inequality

$$\sup_{\mathbf{v}} L(\mathbf{v}, \mathbf{x}^*) \geq L(\mathbf{v}^*, \mathbf{x}^*) \geq \inf_{\mathbf{x}} L(\mathbf{v}^*, \mathbf{x})$$

always holds true.

Thus

$$(\mathbf{v}^*, \mathbf{x}^*) \in \mathbb{R}^{r+m} \times \mathbb{R}^n$$

is a saddle point if and only if

$$\sup_{\mathbf{v}} L(\mathbf{v}, \mathbf{x}^*) = \inf_{\mathbf{x}} L(\mathbf{v}^*, \mathbf{x}).$$

Now, independent of the choice of $\mathbf{x}^*$, it holds true that

$$\begin{aligned} \sup_{\mathbf{v}} L(\mathbf{v}, \mathbf{x}^*) &= \sup_{\mathbf{v}} \left\{ f(\mathbf{x}^*) + v_1 f_1(\mathbf{x}^*) + \cdots + v_{r+m} f_{r+m}(\mathbf{x}^*) \right\} \\ &= f(\mathbf{x}^*) + \delta(\mathbf{x}^* | C_0) > -\infty \end{aligned}$$

where

$$C_0 \subseteq \mathbb{R}^n$$

is the set of all feasible solutions and

$$f(\mathbf{x}) + \delta(\mathbf{x}|C_0) = \begin{cases} f(\mathbf{x}) & \text{if } \mathbf{x} \in C_0 \\ +\infty & \text{if } \mathbf{x} \notin C_0. \end{cases}$$

On the other hand, for each given $\mathbf{v} \in \mathbb{R}^{r+m}$, it holds true that

$$+\infty > \inf_{\mathbf{x}} L(\mathbf{v}, \mathbf{x}) = \begin{cases} \inf h \text{ if } \mathbf{v} \in \mathbb{R}^{r+m}_{\geq 0} \\ -\infty \text{ if } \mathbf{v} \notin \mathbb{R}^{r+m}_{\geq 0} \end{cases}$$

where

$$h(\mathbf{x}) = f_0(\mathbf{x}) + v_1 f_1(\mathbf{x}) + \cdots + v_{r+m} f_{r+m}(\mathbf{x}).$$

Thus $(\mathbf{v}^*, \mathbf{x}^*)$ is a saddle point of $L(\mathbf{v}, \mathbf{x})$ if and only if

$$\mathbf{v}^* \in \mathbb{R}^{r+m}_{\geq 0},\ \mathbf{x}^* \in C_0 \text{ and } \inf h(\mathbf{x}^*) = f(\mathbf{x}^*). \tag{10.4}$$

The condition given in (10.4) is satisfied when $\mathbf{v}^*$ is a KKT vector and $\mathbf{x}^*$ is an optimal solution. Indeed, in that case

$$\inf h(\mathbf{x}^*) = f(\mathbf{x}^*) = \inf f(\mathbf{x}^*) < \infty.$$

On the other hand, suppose that the condition given in (10.4) is satisfied. Then for all $\mathbf{x} \in C_0$, it holds true that

$$v_i f_i(\mathbf{x}) \leq 0$$

for $i = 1, \ldots, r + m$. Therefore

$$h(\mathbf{x}) \leq f(\mathbf{x}).$$

Thus

$$\inf h(\mathbf{x}) = \inf_{\mathbf{x} \in C_0} h(\mathbf{x}) \leq \inf_{\mathbf{x} \in C_0} f(\mathbf{x}) \leq f(\mathbf{x}^*)$$

whence

$$\inf h(\mathbf{x}) = \inf f(\mathbf{x}).$$

Thus (10.4) implies that $\mathbf{v}^*$ is a KKT vector and $\mathbf{v}^*$ is an optimal solution.

Arguing in a similar way, it follows that

$$\inf h(\mathbf{x}) \leq h(\mathbf{x}^*) \leq f(\mathbf{x}^*)$$

for $\mathbf{v} \in \mathbb{R}^{r+m}_{\geq 0}$ and $\mathbf{x} \in C_0$. Here, the second inequality is strict unless

$$v_i f_i(\mathbf{x}) = 0$$

for $i = 1, \ldots, r$.

Thus not only does the condition given by (10.4) imply items (10.1) and (10.2) but also it implies that

$$h(\mathbf{x}^*) = \inf h(\mathbf{x}). \tag{10.5}$$

Conversely, items (10.1), (10.2) and (10.5) imply the condition given by (10.4). Indeed, in this case

$$h(\mathbf{x}^*) = f(\mathbf{x}^*).$$

To complete the proof of the theorem, it remains to show that the condition given by (10.5) is equivalent to the condition given by (10.3) when

$$\mathbf{v} \in \mathbb{R}^{r+m}_{\geq 0}.$$

To this end, by definition, $h(\mathbf{x})$ attains its infimum at $\mathbf{x}^*$ if and only if $\mathbf{0}$ is a subgradient of $h(\mathbf{x})$ at $\mathbf{x}$. Equivalently,

$$\mathbf{0} \in \partial h(\mathbf{x}^*).$$

Now, by assumption

$$\varnothing \neq \bigcap_{i=0}^{r+m} \mathrm{r.\,i.}(\mathrm{dom}(f_i)) = \mathrm{r.\,i.}(C)$$

whence, by Theorem 10.8, it holds true that

$$\partial h(\mathbf{x}) = \partial f(\mathbf{x}) + \partial(v_1 f_1)(\mathbf{x}) + \cdots + \partial(v_{r+m} f_{r+m})(\mathbf{x}) \text{ for all } \mathbf{x}.$$

This condition

$$\mathbf{0} \in \partial h(\mathbf{x})$$

is equivalent to the condition that is given in item (10.3).

Finally, using the assumption of differentiability, by item (ii) in Theorem 10.9, it follows that

$$\partial f(\mathbf{x}) = \nabla f(\mathbf{x}) \text{ and } \partial f_i(\mathbf{x}) = \nabla f_i(\mathbf{x}) \text{ for } i = 1, \ldots, r+m$$

whence the *gradient equation*

$$\nabla f(\mathbf{x}) + v_1 \nabla f_1(\mathbf{x}) + \cdots + v_{r+m} \nabla f_{r+m}(\mathbf{x}) = \mathbf{0}$$

is equivalent to the condition given in item (10.3). □

Exercises

10.1 Consider the following minimization programming problem.

- Minimize the objective function

$$f(x_1, x_2) = -12x_1 + x_1^2 - 50x_2 + x_2^2$$

- Subject to the constraints

$$\begin{aligned} x_1 \quad &\leq 10 \\ x_2 &\leq 15 \end{aligned}$$

and

$$x_1, x_2 \geq 0.$$

(i) Explain why the above programming problem is an ordinary differential convex programming problem.
(ii) Derive the ordinary differential programming problem's Lagrangian.
(iii) Use the KKT conditions to determine an optimal solution.

10.2 Consider the following minimization programming problem.

- Minimize the objective function

$$f(x_1, x_2) = \ln(x_1 + 1) - x_2^2$$

- Subject to the constraints

$$x_1 + 2x_2 \leq 3$$

and

$$x_1, x_2 \geq 0.$$

(i) Explain why the above programming problem is an ordinary differential convex programming problem.
(ii) Derive the ordinary differential programming problem's Lagrangian.
(iii) Use the KKT conditions to determine an optimal solution.

10.3 Consider the following nonlinear maximization programming problem.

- Maximize the objective function

$$f(x_1, x_2) = 3.6x_1 - 0.4x_1^2 + 1.6x_2 - 0.2x_2^2$$

- Subject to the constraints
  - $2x_1 + x_2 \leq 10$ and
  - $x_1, x_2 \geq 0$.

(i) Explain why $f(x_1, x_2)$ is concave.
(ii) Describe the KKT conditions that correspond to the given nonlinear programming problem.
(iii) Determine an optimal solution to the given nonlinear programming problem by solving the corresponding KKT condition equations.

10.4 Show that the following ordinary differential convex programming problem

- Minimize the objective function

$$f(x_1, x_2) = x_1$$

- Subject to the constraints

$$f_1(x_1, x_2) = x_2 \leq 0$$

and

$$f_2(x_1, x_2) = x_1^2 - x_2 \leq 0$$

does not admit a KKT vector. To what extent does this conflict with the conclusion of Theorem 10.11? Explain.

10.5 Define the *indicator function* of $C \subseteq \mathbb{R}^n$ to be

$$\delta(x; C) := \begin{cases} 0 & \text{if } x \in C \\ \infty & \text{otherwise.} \end{cases}$$

Show that C is convex if and only if $\delta(\cdot; C)$ is convex.

10.6 Let $f\colon \mathbb{R}^n \to \mathbb{R} \bigcup \{\pm\infty\}$ be a convex function. Let $a \in \mathbb{R}$. Show that the *level set*

$$\{x \in \mathbb{R}^n : f(x) \leq a\} \subseteq \mathbb{R}^n$$

is convex.

10.7 Let $C \subseteq \mathbb{R}^{n\times n}$ be the collection of $n \times n$ positive definite matrices

$$C := \left\{A = [a_{ij}] \in \mathbb{R}^{n\times n} : \text{ A is positive definite}\right\}.$$

Show that the function

$$\ln(\det(\cdot))\colon C \to \mathbb{R},\ A \mapsto \ln(\det(A))$$

is convex.

10.8 Show that the function

$$f\colon \mathbb{R}^n \to \mathbb{R}\bigcup\{\pm\infty\}$$

is convex if and only if its restriction to each line $\ell \subseteq \mathbb{R}^n$ is convex.

Chapter 11
Using Computer Software to Solve Selected Nonlinear Programming Problems

Overview

- Solving selected nonlinear programming problems using Python.

11.1 Solving Nonlinear Programming Problems Graphically Using Python

Here, we illustrate how to use Python to solve a collection nonlinear convex programming problems. Our approach is graphical and makes use of the Python matplotlib package. More details about this package can be found at

https://matplotlib.org/stable/

Our graphical approach serves to emphasize the key underlying mathematical ideas which surround the theoretical tools and practical methods from Chapters 9 and 10.

While there are several commercially available solvers for solving nonlinear programming problems we do not discuss any such solvers here. Instead, we place emphasis on practical illustrative considerations and graphical techniques.

Example 11.1. Consider the following nonlinear maximization convex programming problem.

- Maximize the objective function

$$Z = \ln(x + 1) - y^2$$

- Subject to the constraints

$$x + 2y \leq 3$$

N. Grieve, *An Introduction to Mathematical Programming and Network Science*,
Springer Undergraduate Texts in Mathematics and Technology,
https://doi.org/10.1007/978-3-032-13330-4_11

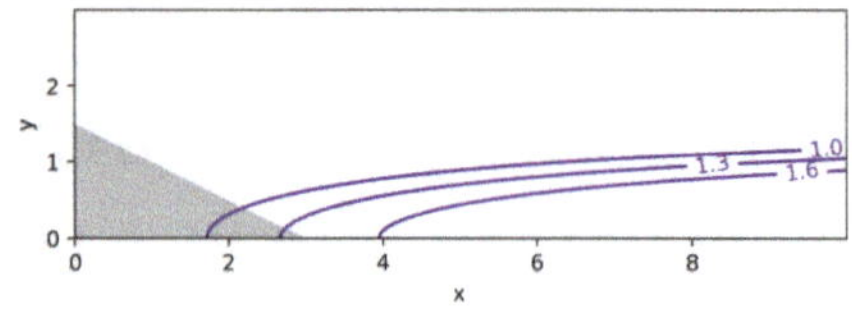

Fig. 11.1 Contours and feasible region for Example 11.1

and

$$x, y \geq 0.$$

This programming problem has a nonlinear convex objective function and the feasible region is cut out by a collection of linear constraints.

Using the matplotlib and numpy packages in Python, the conclusion is that the optimal value of the objective function is $Z = 1.386$. It is achieved at the feasible region's extreme point $(x, y) = (3, 0)$. For example, this can be seen in Figure 11.1 below. It gives a contour diagram of the objective function super imposed onto the feasible region.

Example 11.2. Consider the nonlinear minimization convex programming problem

- Minimize the objective function

$$Z = x^2 - 6x + y^3 - 3y$$

- Subject to the constraints

$$x + y \leq 1$$

and

$$x, y \geq 0.$$

Using the mathplolib package in Python, this programming problem can be solved by studying the contours of the objective function super imposed onto the feasible region. This is illustrated in Figure 11.2.

The conclusion is that the optimal value of the objective function is

$$Z = -5.$$

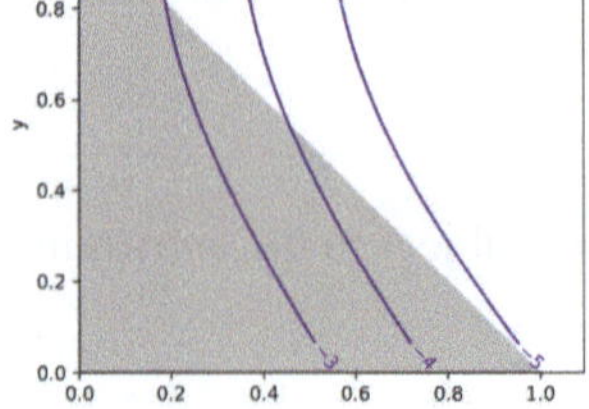

Fig. 11.2 Contours and feasible region for Example 11.2

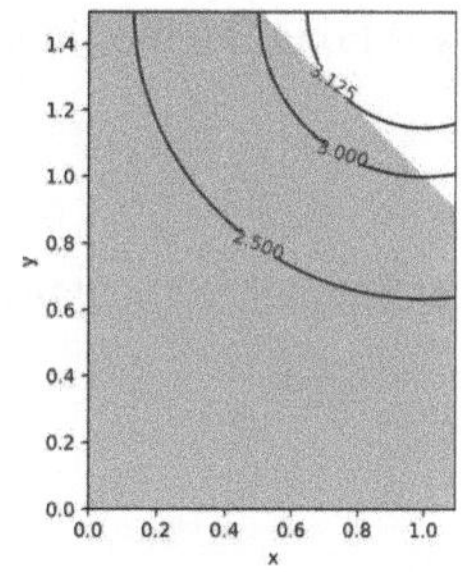

Fig. 11.3 Contours and feasible region for Example 11.3

It is found at the feasible region's extreme point $(x, y) = (1, 0)$.

Example 11.3. The quadratic maximization programming problem

- Maximize the quadratic objective function

$$Z = 2x + 3y - x^2 - y^2$$

- Subject to the constraints

$$x + y \leq 2$$

and

$$x, y \geq 0$$

can be solved by studying the contours of the objective function superimposed onto the feasible region. (See Figure 11.3.)

The conclusion is that the optimal value of the objective function is

$$Z = 3.125.$$

It is achieved at the point $(0.75, 1.25)$.

11.2 Solving Separable Programming Problems Using Python

Example 11.4. Here we revisit Example 9.22 and show how it can be solved using the Python scipy.optimize package. Indeed, recall that the given separable programming problem, which has nonlinear separable objective function

$$f(x_1, x_2) = f_1(x_1) + f_2(x_2)$$

where $y_1 = f_1(x_1)$ and $y_2 = f_2(x_2)$ are, respectively, defined by the graphs given in Figure 9.9, is modelled by the maximization linear programming problem

- Maximize the linear objective function

$$Z = 3x_{11} + 2x_{12} + 5x_{21} + x_{22}$$

- Subject to the constraints

$$x_{11} + x_{21} \leq 4$$

$$2(x_{21} + x_{22}) \leq 12$$

$$3(x_{11} + x_{12}) + 2(x_{21} + x_{22}) \leq 18$$

$$x_{11} \leq 3, x_{12} \leq 1, x_{21} \leq 3, x_{22} \leq 3$$

and

$$x_{11} \geq 0, x_{12} \geq 0, x_{21} \geq 0, x_{22} \geq 0.$$

This linear programming problem can be solved using scipy.optimize in the following way.

The Python scipy.optimize scripts for Example 11.4

```
>>> from scipy.optimize import linprog
... obj_func = [-3,-2,-5,-1]
... lhs_ineq = [[1,0,1,0],
...             [0,0,2,2],
...              [3,3,2,2],
...              [1,0,0,0],
...              [0,1,0,0],
...              [0,0,1,0],
...              [0,0,0,1]]
... rhs_ineq = [4,12,18,3,1,3,3]
... opt_soltn = linprog(c=obj_func, A_ub=lhs_ineq, b_ub=rhs_ineq, method="highs")
... print(opt_soltn)
       message: Optimization terminated successfully. (HiGHS Status 7: Optimal)
       success: True
        status: 0
           fun: -23.0
             x: [ 1.000e+00  1.000e+00  3.000e+00  3.000e+00]
           nit: 2
         lower:  residual: [ 1.000e+00  1.000e+00  3.000e+00  3.000e+00]
                marginals: [ 0.000e+00  0.000e+00  0.000e+00  0.000e+00]
         upper:  residual: [       inf        inf        inf        inf]
                marginals: [ 0.000e+00  0.000e+00  0.000e+00  0.000e+00]
         eqlin:  residual: []
                marginals: []
       ineqlin:  residual: [ 0.000e+00  0.000e+00  0.000e+00  2.000e+00
                             0.000e+00  0.000e+00  0.000e+00]
                marginals: [-1.500e+00 -0.000e+00 -5.000e-01 -0.000e+00
                            -5.000e-01 -2.500e+00 -0.000e+00]
 mip_node_count: 0
 mip_dual_bound: 0.0
        mip_gap: 0.0
```

Exercises

11.1 Use a contour diagram plot to estimate a local minimum of

$$f(x_1, x_2) = x_1^2 + 3x_2^2.$$

Is this minimum a global minimum? Why or why not?

11.2 Use a contour diagram to solve the nonlinear programming problem

- Maximize the objective function

$$f(x_1, x_2) = 3.6x_1 - 0.4x_1^2 + 1.6x_2 - 0.2x_2^2$$

- Subject to the constraints
  - $2x_1 + x_2 \leq 10$ and
  - $x_1, x_2 \geq 0$.

11.3 Use a contour diagram to estimate an optimal solution of the nonlinear programming problem

- Minimize the objective function

$$f(x_1, x_2) = -12x_1 + x_1^2 - 50x_2 + x_2^2$$

- Subject to the constraints

$$\begin{aligned} x_1 \quad &\leq 10 \\ x_2 &\leq 15 \end{aligned}$$

and

$$x_1, x_2 \geq 0.$$

11.4 Consider the case that a company seeks to determine the best product mix for two new products. Since the products would share the same production facilities, the total number of units produced of the two products combined cannot exceed two per hour. On the other hand, because of the uncertainty of how well these products will sell, the profit from producing each product provides a decreasing marginal return as the production rate is increased. Now, given a production rate of R_1 units per hour, it is estimated that Product 1 would provide a profit per hour of

$$\$200R_1 - \$100R_1^2;$$

if the product rate of Product 2 is R_2 units per hour, then its estimated profit per hour would be

$$\$300R_2 - \$100R_2^2.$$

(i) Formulate a quadratic programming model for determining the product mix that maximizes the total profit per hour.
(ii) Use a contour plot to solve the corresponding quadratic programming problem.

Chapter 12
Introduction to Game Theory, Decision Analysis and Multicriteria Optimization

Overview

- Matrix games and linear programming.
- Introduction to decision analysis.
- Introduction to multicriteria optimization.

12.1 Game Theory and Matrix Games

In a *matrix game*, there is an $m \times n$ *payoff matrix*, a *row player* and a *column player*. In each round of the game, the *row player chooses* a *row* of the payoff matrix and the *column player* chooses a *column* of the *payoff matrix*. These choices are then cross-indexed to find the *payoff* for the round of the game.

Our conventions, in what follows, are such that the *payoffs in the matrix are always listed as winnings for the row player*. So, winnings for the column player appear as losses for the row player. In a matrix game, a row player aims to maximize the payoff; the column player wishes to minimize the payoff. The framework of matrix games encompasses the idea of *zero sum game*. The idea is that one player's loss will result in another player's gain.

An *optimal strategy* for a player of a matrix game, is a strategy in which a player can maximize the winnings or minimize the losses—this is all under the assumption that the other player will have perfect knowledge of this strategy and will also play so as to maximize winnings or minimize losses subject to this strategy. In other words, an optimal strategy for a player in a matrix game assumes an all knowing opponent.

N. Grieve, *An Introduction to Mathematical Programming and Network Science*,
Springer Undergraduate Texts in Mathematics and Technology,
https://doi.org/10.1007/978-3-032-13330-4_12

12.1.1 Probabilistic Strategies for Matrix Games

For our purposes, a *matrix game* is an $m \times n$ real matrix

$$A = [a_{ij}] \in \mathbb{R}^{m \times n}.$$

The entries a_{ij} of the matrix represent the *payoffs*, respectively, *loses*, for the row player, respectively, the column player, when the *pure row strategy* i and the *pure column strategy* j is chosen.

A *mixed probabilistic strategy* for the row player is a column vector

$$\mathbf{p} = \begin{bmatrix} p_1 \\ \vdots \\ p_m \end{bmatrix} \in \mathbb{R}^{m \times 1}$$

which is such that

(i) $p_i \geq 0$ for all $i = 1, \ldots, m$; and
(ii) $\sum_{i=1}^{m} p_i = 1$.

A *mixed probabilistic strategy* for the column player is a row vector

$$\mathbf{q} = \begin{bmatrix} q_1 & \ldots & q_n \end{bmatrix} \in \mathbb{R}^{1 \times n}$$

which is such that

(i) $q_j \geq 0$ for all $j = 1, \ldots, n$; and
(ii) $\sum_{j=1}^{n} q_j = 1$.

Pure strategies are those mixed strategies which contain a single entry which equals 1 (and thus have all other entries equal to zero). The idea of mixed strategies **p** and **q**, respectively, is that the row player should employ the row strategy i, for $i = 1, \ldots, m$, with probability p_i, whereas the column player should employ the column strategy j, for $j = 1, \ldots, n$, with probability q_j.

12.1.2 Expected Winnings and Mixed Strategies for Matrix Games

If the column player chooses the column j pure strategy, for $j = 1, \ldots, n$, then the expected winnings for the row player when using a mixed strategy

$$\mathbf{p} = \begin{bmatrix} p_1 \\ \vdots \\ p_m \end{bmatrix}$$

is given by

$$E_j(\mathbf{p}) = p_1 a_{1j} + \cdots + p_m a_{mj}$$

for $j = 1, \ldots, n$.

Since the row player aims to ensure that the minimum expected winnings are as large as possible, the optimal strategy for the row player is to choose the mixed strategy $\mathbf{p}$ which is such that the quantity

$$\min_{1 \le j \le n} E_j(\mathbf{p})$$

is maximal.

Similarly, if the row player chooses the row i pure strategy, for $i = 1, \ldots, m$, and if the column player uses the mixed strategy

$$\mathbf{q} = \begin{bmatrix} q_1 & \ldots & q_n \end{bmatrix}$$

then the expected value of the losses is given by

$$F_i(\mathbf{q}) = q_1 a_{i1} + \cdots + q_n a_{in}$$

for $i = 1, \ldots, m$.

Since the column player aims to ensure that the maximum expected losses are as small as possible, the optimal strategy for the column player is to choose the mixed strategy $\mathbf{q}$ which is such that

$$\max_{1 \le m} F_i(\mathbf{q})$$

is minimal.

To summarize the above discussion, the conclusion is that the expected winnings of the row player for each round of the game—provided that both players play their optimal strategies—is given by

$$u := \max_{\mathbf{p}} \min_{1 \le j \le n} E_j(\mathbf{p});$$

the expected winnings of the column player for each round of the game is given by

$$v := \min_{\mathbf{q}} \max_{1 \le i \le m} F_i(\mathbf{q}).$$

As a final comment about *optimality* of strategies for the row and column player, respectively, it is important to emphasize the inherit, and in some sense subjective, assumptions about what factors may influence the row and column player's respective *utility function*.

For instance, until specifying in precise terms the extent to which the row and column player, respectively, may or may not be risk-neutral, risk-seeking or risk-averse, there is no way to assert definitively that any optimal strategy is indeed optimal with respect to a given utility function. We omit a further discussion about such considerations here.

12.1.3 Optimal Strategies for Matrix Games

It turns out that such matrix games can be *solved*, in the sense that the row and column players can find *optimal mixed strategies* via suitably formulated dual noncanonical linear programming problems. Indeed, this is the content of Theorem 12.1 below.

Theorem 12.1. *Let*

$$A = [a_{ij}] \in \mathbb{R}^{m \times n}$$

be an $m \times n$ *matrix game. Let*

$$u := \max_{\mathbf{p}} \min_{1 \leq j \leq n} E_j(\mathbf{p})$$

be the expected winnings of the row player and let

$$v := \min_{\mathbf{q}} \max_{1 \leq i \leq m} F_i(\mathbf{q})$$

be the expected winnings of the column player.

Then u is the optimal value of the maximization linear programming problem

- *Maximize the linear objective function*

$$x_{m+1}$$

- *Subject to the constraints*

$$\begin{array}{ccccccc} a_{11}x_1 & + & \ldots & + & a_{m1}x_n & - & x_{m+1} \geq 0 \\ \vdots & & \vdots & & \vdots & & \vdots \quad \vdots \\ a_{1n}x_1 & + & \ldots & + & a_{mn}x_n & - & x_{m+1} \geq 0 \end{array}$$

$$x_1 + \cdots + x_m = 1$$

and

$$x_1, \ldots, x_m \geq 0.$$

Further, v is the optimal value of the minimization linear programming problem

- *Minimize the linear objective function*

$$y_{n+1}$$

- *Subject to the constraints*

$$\begin{array}{ccccc} a_{11}y_1 & + \ldots + & a_{1n}y_n & - \; y_{n+1} & \leq 0 \\ \vdots & \vdots & \vdots & \vdots & \vdots \\ a_{m1}y_1 & + \ldots + & a_{mn}y_n & - \; y_{n+1} & \leq 0 \end{array}$$

$$y_1 + \cdots + y_n = 1$$

and

$$y_1, \ldots, y_n \geq 0.$$

Proof. We consider the case of the row player. The case of the column player is similar. The aim of the row player is to employ an optimal mixed strategy **p** which is characterized by the condition that it maximizes the minimum expected payoffs.

Introducing decision variables $x_1, \ldots, x_{m+1}$, it follows that the optimal row player mixed strategies **p** are achieved as optimal solutions

$$x_1 = p_1, \ldots, x_m = p_m, x_{m+1} = u$$

of the maximization linear programming problem

- Maximize the linear objective function

$$x_{m+1}$$

- Subject to the constraints

$$\begin{array}{ccccc} a_{11}x_1 & + \ldots + & a_{m1}x_n & \geq & x_{m+1} \\ \vdots & \vdots & \vdots & & \vdots \\ a_{1n}x_1 & + \ldots + & a_{mn}x_n & \geq & x_{m+1} \end{array}$$

$$x_1 + \cdots + x_m = 1$$

and

$$x_1, \ldots, x_m \geq 0.$$

As mentioned, the conclusion for the case of the column player follows similarly. □

Theorem 12.1 is illustrated in Example 12.2 below.

Example 12.2. Consider the matrix game

$$A = \begin{bmatrix} -3 & 4 \\ 2 & -3 \end{bmatrix} = \begin{bmatrix} a_{11} & a_{12} \\ a_{21} & a_{22} \end{bmatrix}.$$

The corresponding maximization and minimization linear programming problems are described in the following way.

- Maximize the linear objective function

$$x_3 = \min_{1 \le j \le 2} E_j(\mathbf{p}) = \min_{1 \le j \le 2} x_1 a_{1j} + x_2 a_{2j}$$

- Subject to the constraints

$$\begin{aligned} -3x_1 + 2x_2 &\ge x_3 \\ 4x_1 - 3x_2 &\ge x_3 \end{aligned}$$

$$x_1 + x_2 = 1$$

  and

$$x_1, x_2 \ge 0;$$

respectively,

- Minimize the linear objective function

$$y_3 = \max_{1 \le i \le 2} F_i(\mathbf{q}) = \max_{1 \le i \le 2} y_1 a_{i1} + y_2 a_{i2}$$

- Subject to the constraints

$$\begin{aligned} -3y_1 + 4y_2 &\le y_3 \\ 2y_1 - 3y_2 &\le y_3 \end{aligned}$$

$$y_1 + y_2 = 1$$

  and

$$y_1, y_2 \ge 0.$$

Solving these respective linear programming problems, the optimal strategy for the row player is

$$\mathbf{p}^* = \begin{bmatrix} p_1 \\ p_2 \end{bmatrix} = \begin{bmatrix} 0.4167 \\ 0.5833 \end{bmatrix}$$

whereas the optimal strategy for the column player is

$$\mathbf{q}^* = \begin{bmatrix} q_1 & q_2 \end{bmatrix} = \begin{bmatrix} 0.5833 & 0.4167 \end{bmatrix}.$$

Further

$$u = \max_{\mathbf{p}} \min_{1 \le j \le 2} E_j(\mathbf{p}) = -0.0833$$

and

$$v = \min_{\mathbf{q}} \max_{1 \le i \le 2} F_i(\mathbf{q}) = -0.0833.$$

In particular, the von Neumann value of the game, in the sense of Theorem 12.3 below, is −0.0833.

The following Python scripts illustrate how to solve the above linear programming problems using the scipy optimize package.

Python scipy.optimize scripts for Example 12.2

```
>>> from scipy.optimize import linprog
... obj_func = [0,0,-1]
... lhs_ineq = [[3,-2,1],[-4,3,1]]
... rhs_ineq = [0,0]
... lhs_eq = [[1,1,0]]
... rhs_eq = [1]
... x1_bounds = (0, None)
... x2_bounds = (0,None)
... x3_bounds = (None,None)
... bounds = [x1_bounds, x2_bounds, x3_bounds]
... opt_soltn = linprog(c=obj_func, A_ub=lhs_ineq,
...                     b_ub=rhs_ineq, A_eq = lhs_eq, b_eq = rhs_eq,
...                     bounds=bounds, method="highs")
... print(opt_soltn)
...
        message: Optimization terminated successfully. (HiGHS Status 7: Optimal)
        success: True
         status: 0
            fun: 0.08333333333333348
              x: [ 4.167e-01  5.833e-01 -8.333e-02]
            nit: 2
          lower:  residual: [ 4.167e-01  5.833e-01        inf]
                 marginals: [ 0.000e+00  0.000e+00  0.000e+00]
          upper:  residual: [       inf        inf        inf]
                 marginals: [ 0.000e+00  0.000e+00  0.000e+00]
          eqlin:  residual: [ 0.000e+00]
                 marginals: [ 8.333e-02]
        ineqlin:  residual: [ 0.000e+00  0.000e+00]
                 marginals: [-5.833e-01 -4.167e-01]
 mip_node_count: 0
 mip_dual_bound: 0.0
        mip_gap: 0.0
>>> obj_func = [0,0,1]
... lhs_ineq = [[-3,4,-1],[2,-3,-1]]
... rhs_ineq = [0,0]
... lhs_eq = [[1,1,0]]
... rhs_eq = [1]
... x1_bounds = (0, None)
... x2_bounds = (0,None)
... x3_bounds = (None,None)
```

```
... bounds = [x1_bounds, x2_bounds, x3_bounds]
... opt_soltn = linprog(c=obj_func, A_ub=lhs_ineq, b_ub=rhs_ineq,
...                     A_eq = lhs_eq, b_eq = rhs_eq,
...                     bounds=bounds, method="highs")
... print(opt_soltn)
...
        message: Optimization terminated successfully. (HiGHS Status 7: Optimal)
        success: True
         status: 0
            fun: -0.08333333333333348
              x: [ 5.833e-01  4.167e-01 -8.333e-02]
            nit: 2
          lower:  residual: [ 5.833e-01  4.167e-01        inf]
                 marginals: [ 0.000e+00  0.000e+00  0.000e+00]
          upper:  residual: [       inf        inf        inf]
                 marginals: [ 0.000e+00  0.000e+00  0.000e+00]
          eqlin:  residual: [ 0.000e+00]
                 marginals: [-8.333e-02]
        ineqlin:  residual: [ 0.000e+00  0.000e+00]
                 marginals: [-4.167e-01 -5.833e-01]
 mip_node_count: 0
 mip_dual_bound: 0.0
        mip_gap: 0.0
```

12.1.4 Additional Perspective for Solving Matrix Games via Linear Programming: The von Neumann Minimax Theorem

We conclude our discussion of matrix games and their solution via linear programming by stating the von Neumann Minimax Theorem for matrix games. (See Theorem 12.3.) This theorem establishes existence of the *von Neumann number* for each given matrix game.

Theorem 12.3 (von Neumann Minimax Theorem). *Let*

$$A = [a_{ij}] \in \mathbb{R}^{m \times n}$$

be an $m \times n$ matrix game. Then there exists optimal mixed strategies $\mathbf{p}^*$ and $\mathbf{q}^*$ for the row and column player, respectively. Further setting

$$u := \max_{\mathbf{p}} \min_{1 \le j \le n} E_j(\mathbf{p})$$

and

$$v := \min_{\mathbf{q}} \max_{1 \le i \le m} F_i(\mathbf{q})$$

it then holds true that

$$\min_{1\leq j\leq n} E_j(\mathbf{p}^*) = u = v \max_{1\leq i\leq m} F_i(\mathbf{q}^*).$$

This common number

$$u = v$$

is called the von Neumann value of the game.

12.2 Selected Introduction to Decision Analysis

In the situation of matrix games, see Section 12.1, an underlying focus is on decision making when the consequences of alternative decisions are known with a reasonable degree of certainty. By this perspective, both the row and column player are able to determine optimal strategies. Here, we describe selected methodologies for decision analysis as an extension of the theory of matrix games.

Example 12.4. Suppose that two companies share the bulk of the market for a particular kind of product. Suppose that each is now making its new marketing plans for the next year in an attempt to take away some sales from the other company.

Then the total sales for the product are relatively fixed, so one company can increase its sales only by taking them away from the other. Suppose that each company is considering three possibilities

- better packaging of the product;
- increased advertising; and
- a slight reduction in price.

The costs of the three alternatives are quite comparable and sufficiently large so that each company will just select one. The estimated effect of each combination of alternatives on the *increased percentage of the sales* for Company 1 is summarized in the following payoff table

$$\begin{bmatrix} 2 & 3 & 1 \\ 1 & 4 & 0 \\ 3 & -2 & -1 \end{bmatrix} = \begin{bmatrix} a_{11} & a_{12} & a_{13} \\ a_{21} & a_{22} & a_{23} \\ a_{31} & a_{23} & a_{33} \end{bmatrix}.$$

View Company 1 as the *row player* and Company 2 as the *column player*. Then each company, viewed as a player in the matrix game, must make its selection before learning the decision of the other company.

Within the above set-up, Company 1 would find the optimal mixed strategy by using the Simplex Method to solve the following linear programming problem.

- Maximize the linear objective function

$$f := x_4$$

- Subject to the constraints

$$\begin{aligned}
a_{11}x_1 + a_{21}x_2 + a_{31}x_3 - x_4 &\geq 0 \\
a_{12}x_1 + a_{22}x_2 + a_{32}x_3 - x_4 &\geq 0 \\
a_{13}x_1 + a_{23}x_2 + a_{33}x_3 - x_4 &\geq 0
\end{aligned}$$

$$x_1 + x_2 + x_3 + x_4 = 1$$

and

$$x_i \geq 0 \text{ for } i = 1, 2, 3.$$

Similarly, Company 2 would find the optimal mixed strategy by using the Simplex Method to solve the following linear programming problem.

- Minimize the linear objective function

$$g := y_4$$

- Subject to the constraints

$$\begin{aligned}
a_{11}y_1 + a_{12}y_2 + a_{13}y_3 - y_4 &\leq 0 \\
a_{21}y_1 + a_{22}y_2 + a_{23}y_3 - y_4 &\leq 0 \\
a_{31}y_1 + a_{32}y_2 + a_{33}y_3 - y_4 &\leq 0
\end{aligned}$$

$$y_1 + y_2 + y_3 + y_4 = 1$$

and

$$y_i \geq 0 \text{ for } i = 1, 2, 3.$$

Recall, that the *minimax theorem* (Theorem 12.3) then says that within the presence of mixed strategies, the pair of mixed strategies that is optimal according to the minimax criterion provides a *stable solution* so that neither player can do better by unilaterally changing strategy.

By contrast to matrix games, what we want to consider now is a decision making framework when the consequences of alternative decisions are not known with a large degree of certainty.

12.2.1 Key Principles and Guiding Questions for Decision Analysis

In what follows below are a few guiding examples which illustrate the main guiding question for decision analysis. For instance, we describe decision making methodologies which admit applications to situations which present a great amount of uncertainty in terms of consequences for deciding on any particular course of action. Some representative situations are described in Examples 12.5, 12.6 and 12.7.

Example 12.5. Suppose that a manufacturer introduces a new product into the market place.

- What will be the reaction of potential customers?
- How much should be produced?
- How much advertising is needed to launch the product successfully?

Example 12.6. Suppose that an agricultural firm is selecting the mix of crops and livestock for the upcoming season.

- What will be the weather conditions?
- Where are prices headed?
- What will the costs be?

Example 12.7. Suppose that an oil company will decide whether or not to drill for oil in a particular location.

- How likely is oil there?
- How much oil is there?
- Should geologists investigate the site further prior to deciding about drilling?

As these motivational examples suggest, some typical first questions to consider when considering decision making methodologies are

- If the needed decisions should be made immediately; and/or
- If first some *testing* (and at some expense) should be done in order to reduce the level of uncertainty about the given situation.

In what follows, we will want to explore techniques for making informed decisions within the presence of uncertainty for the consequences of such decisions.

Table 12.1 Resource extraction table for Example 12.8

	Contains natural resource	Does not contain natural resource
Extract resource	\$70, 000, 000	−\$10, 000, 000
Sell the land	\$9, 000, 000	\$9, 000, 000
Chance of success	25%	75%

12.2.2 A Methodology for Decision Making

Here, we describe a methodology for decision making. In general terms, the decision maker must make a choice from a set of possible decisions. The idea is that the set contains all of the *feasible choices* to consider for the given problem at hand. There are also *random factors* that determine various situations.

These considerations influence the decision making process. The *payoff* is a quantitative measure of the value to the decision maker of the consequences of any given outcome. A *payoff table* can be oftentimes used to provide the payoff for each combination of an action and the random factor.

Example 12.8 (Resource extraction). Suppose that Company A owns land that might contain an abundance of a certain natural resource. Suppose that a consulting geologist has reported to management that there is a 25% chance that this natural resource exists. Suppose that Company B has offered to purchase the land for \$9, 000, 000. However, Company A is considering holding on to the land in order to extract the resource.

Suppose that the projected cost of extracting the resource is \$10, 000, 000. Further, if the natural resource is found, suppose that the resulting expected revenue will be \$8, 000, 000. Thus, the company's expected profit after deducing the cost of extraction is \$70, 000, 000. A loss of \$10, 000, 000 (the extraction cost) will thus be incurred if the land does not contain the natural resource.

The relevant data is summarized in Table 12.1. This is a *prospective profits payoff table*.

In general, the decision maker and the random factors can be viewed as two players in a game. In this context, a *decision analysis framework* is formulated in Definition 12.9 below.

Definition 12.9 (Decision analysis framework).

- The decision maker chooses a decision option.
- A random factor (which is out of control of the decision marker) will occur.
- Each combination of a decision option and random factor results in a payoff which is given as one of the entries in a payoff table.
- This payoff table should be used to find an optimal choice for the decision maker according to an appropriate criterion.

Table 12.2 Decision analysis table for Example 12.10

	Contains natural resource	Does not contain natural resource
Extract resource	$70, 000, 000	−$10, 000, 000
Sell the land	$9, 000, 000	$9, 000, 000
Prior probability	.25	.75

When comparing the general decision framework to the more traditional game theory context, there are some important considerations to keep in mind. For example, when considering the case of matrix games, both players are assumed to be rational and thus choosing their strategies to promote their own benefit. In the more general decision making context, the decision maker can be assumed to be rational and choosing strategies to promote their own welfare; but random factors are considered as a passive player and each random factor arises randomly as a chosen strategy.

Another aspect to the general decision making framework is the concept of *prior distribution*. The idea here, is that the decision maker has some information about the relative likelihood of each of the possible random factors. The probabilities for respective states of nature can be called the *prior probabilities*.

Example 12.10. Keeping with the situation that we considered in Example 12.8, the payoff table with prior probabilities is given in Table 12.2.

In what follows, we describe criteria for making decisions based on this data.

12.2.3 Decision Analysis Maximum Payoff Criterion

The *maximum payoff criterion* for decision analysis is formulated in the following way.

Definition 12.11 (Maximum payoff criterion)**.** For each possible decision alternative, find the *minimum payoff* over all possible random factors. Next, find the *maximum* of these minimum payoffs and choose the alternative whose minimum payoff gives this maximum.

In practice the maximum payoff criterion should be viewed as a fairly conservative strategy.

Example 12.12. If we apply the maximum payoff criterion to the payoff table given in Table 12.2, then the decision taken will be to sell the land since the maximin value is $9, 000, 000. The rationale is that selling the land provides the best guarantee of the payoff that will be obtained. The criterion essentially takes the pessimistic viewpoint that regardless of which alternative is selected, the worst random factor for that alternative is likely to occur; so the decision should be to choose the alternative which provides the best payoff with the worst random factor.

12.2.4 Decision Analysis Maximum Likelihood Criterion

The decision analysis *maximum likelihood criterion* is formulated as follows.

Definition 12.13 (Maximum likelihood criterion)**.** Identify the most likely random factor—that is, the one with the largest prior probability. For this particular random factor, find the decision alternative with the maximum payoff. Choose this decision alternative.

Example 12.14. If we apply the *maximum likelihood criterion* to the payoff table given in Table 12.2, then the observation is that the outcome of not containing the natural resource has the largest prior probability. In this column, the sell alternative has the maximum payoff, so the choice is to sell the land. The appeal of the maximum likelihood criterion is that the most important random factor is the most likely one. So, the alternative chosen is the best one for this apparently likely random factor.

Basing the decisions on the assumption that the most likely random factor will occur tends to give a better chance of a favourable outcome compared to assuming any other random factor. Another advantage, is that the criterion does not rely on questionable subjective estimates of the probabilities of the respective states of nature other than identifying the mostly likely factor.

One drawback to the maximum likelihood criterion is that it ignores much relevant information. No random factor is considered other than the most likely one. In effect, the criterion does not permit gambling on a low-probability big payoff, no matter how attractive the gamble may be.

12.2.5 Bayes' Decision Rule

Bayes' Rule for decision analysis is formulated in the following way.

Definition 12.15 (Bayes' Rule)**.** Using the best available estimates of the probabilities of the respective random factors, that is the prior probabilities, calculate the expected value of the payoff for each of the possible decision alternatives. Choose the decision alternative that yields the maximum expected payoff.

Example 12.16. For the payoff table given in Table 12.2, we calculate that

- the expected payoff by extracting the resource is equal to

$$(0.25)(70,000,000) + (0.75)(-10,000,000) = 1.0 \times 10^7;$$

  whereas
- the expected payoff by selling the land is equal to

$$(0.25)(9,000,000) + (0.75)(9,000,000) = 9.0 \times 10^6.$$

Finally, since

$$1.0 \times 10^7 > 9.0 \times 10^6$$

the conclusion by employing Bayes' Rule for decision analysis is to extract the resource.

An advantage to Bayes' decision rule is that it incorporates all of the available information. This includes all of the payoffs and all of the best available estimates of the probabilities of the respective states. However, a drawback is that the estimates of the probabilities are perhaps subjective and thus might be unreliable.

There is no accurate way to predict the future—even in probabilistic terms. On the other hand, in many situations reasonable probabilistic models exist. So, Bayes' decision rule is a good criterion to employ in many situations given sufficiently solid probability models.

12.2.6 Decision Making with Experimentation and Conditional Probabilities

Oftentimes, the preliminary estimates of the probabilities that arise within a given payoff table type model can be improved with additional testing and experimentation. These improved estimates are called *posterior probabilities*.

To place matters into perspective, we first recall the statement of Bayes' Theorem from Probability Theory (see Theorem 12.17 below). Recall, that this theorem is deduced via the *total law of probability*.

Before stating Theorem 12.17, recall that a collection of events are *mutually exclusive* if no two have a common outcome. Further, a collection of events are *mutually exhaustive* if at least one must occur. With these preliminaries at hand, Bayes' Theorem is formulated in the following way.

Theorem 12.17 (Bayes' Theorem). *Let $A_1, \ldots, A_k$ be a collection of mutually exclusive and mutually exhaustive events and having* prior *probabilities $P(A_i)$ for $i = 1, \ldots, k$. Then, for each other event B with $P(B) > 0$, the* posterior *probability of A_j, given that B has occurred is given by*

$$P(A_j|B) = \frac{P(A_j \bigcap B)}{P(B)} = \frac{P(B|A_j)P(A_j)}{\sum_{i=1}^{k} P(B|A_i)P(A_i)}$$

for $j = 1, \ldots, k$.

Having recalled the statement of Bayes' Theorem, we now revisit Example 12.8.

Example 12.18. Consider again the payoff table given in Table 12.2. To get a better estimate as to the probability of the abundance of the natural resource that Company

A wishes to extract, suppose that it is possible to conduct a detailed geological survey of the land at a cost of \$300, 000.

To take this possible additional information into the decision making process, the idea is to construct a suitable a probability model. This will provide mathematically grounded insight to take into account upon considering the decision to proceed with the resource extraction compared to taking the decision to sell the land.

We first consider the possible outcomes of the geological survey.

- Unfav.sur. = Unfavourable geological survey—abundance of the natural resource is fairly unlikely.
- Fav.sur. = Favourable geological survey—abundance of the natural resource is fairly likely.

Suppose that past experience shows that if there is an abundance of the natural resource, then the probability of an unfavourable geological survey is

$$P(\text{Unfav.sur.}|\text{Abund. Res.}) = 0.4.$$

It follows that if there is an abundance of the natural resource, then the probability of a favourable geological survey is

$$P(\text{Fav.sur.}|\text{Abund. Res.}) = 1 - 0.4 = 0.6.$$

Similarly, suppose that based on past experience it is known that the probability of unfavourable geological survey given a lack of resources is estimated to be

$$P(\text{Unfav.sur.}|\text{Not Abund. Res.}) = 0.8.$$

It follows that if there is a lack of resources, then the probability of a favourable geological survey is

$$P(\text{Fav.sur.}|\text{Not Abund. Res.}) = 1 - 0.8 = 0.2.$$

We can then calculate the *posterior probabilities* using *Bayes' Theorem*. (See Theorem 12.17.) Indeed, in light of the conditional probability data given above the posterior probabilities are calculated in the following way.

$$P(\text{Abund. Res.}|\text{Unfav.sur.}) = \frac{(0.4)(0.25)}{(0.4)(0.25) + (0.8)(0.75)} = \frac{1}{7}$$

$$P(\text{Not Abund. Res.}|\text{Unfav.sur.}) = 1 - P(\text{Abund. Res.}|\text{Unfav.sur.}) = 1 - \frac{1}{7} = \frac{6}{7}$$

$$P(\text{Abund. Res.}|\text{Fav.sur.}) = \frac{(0.6)(0.25)}{(0.6)(0.25) + (0.2)(0.75)} = \frac{1}{2}$$

Table 12.3 Decision analysis table for Example 12.18

Finding from surv.	Opt. Dec.	Exp. payoff exclud. surv. cost	Exp. payoff including surv. cost
Unfav. sur.	Sell land	$\$9.0 \times 10^6$	$\$8.7 \times 10^6$
Fav. sur.	Extract Res.	$\$3.0 \times 10^7$	$\$2.97 \times 10^7$

$$P(\text{Not Abund. Res.}|\text{Fav.sur.}) = 1 - P(\text{Abund. Res.}|\text{Fav.sur.}) = 1 - \frac{1}{2} = \frac{1}{2}.$$

We now consider *expected payoff values*. First, we need to calculate the following conditional expected values.

$$E(\text{Extract Res.}|\text{Unfav.sur.}) = \frac{1}{7}(70,000,000) + \frac{6}{7}(-10,000,000) = 1.4286 \times 10^6$$

$$E(\text{Sell land}|\text{Unfav.sur.}) = \frac{1}{7}(9,000,000) + \frac{6}{7}(9,000,000) = 9.0 \times 10^6$$

$$E(\text{Extract Res.}|\text{Fav.sur.}) = \frac{1}{2}(70,000,000) + \frac{1}{2}(-10,000,000) = 3.0 \times 10^7$$

$$E(\text{Sell land}|\text{Fav.sur.}) = \frac{1}{2}(9,000,000) + \frac{1}{2}(9,000,000) = 9.0 \times 10^6.$$

We then subtract off from these expected values the \$300,000 cost of conducting the geological survey.

Now, since the objective is to maximize the expected payoff, the above calculations can be summarized in a *optimal policy recommendation table*. (See Table 12.3.)

To summarize, the *conditional probability tree diagram* given in Figure 12.1 represents the steps involved in applying conditional probability informed decision rules.

12.2.7 Decision Trees as a Tool for Decision Analysis

As a concluding conceptual summary for the main steps involved using decision analysis to make informed decisions, the *decision tree diagram* given in Figure 12.2 summarizes the possible situations that can arise in the resource extraction example that we have been considering.

Given such decision trees, calculating the total payoff for each possible outcome provides an informed methodology for making informed decisions.

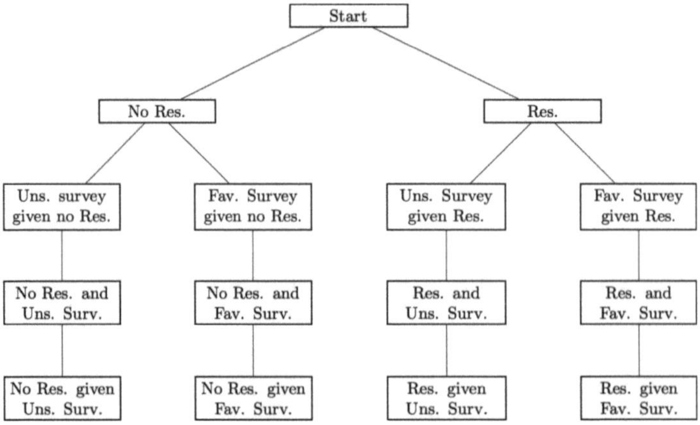

Fig. 12.1 Conditional probability tree diagram

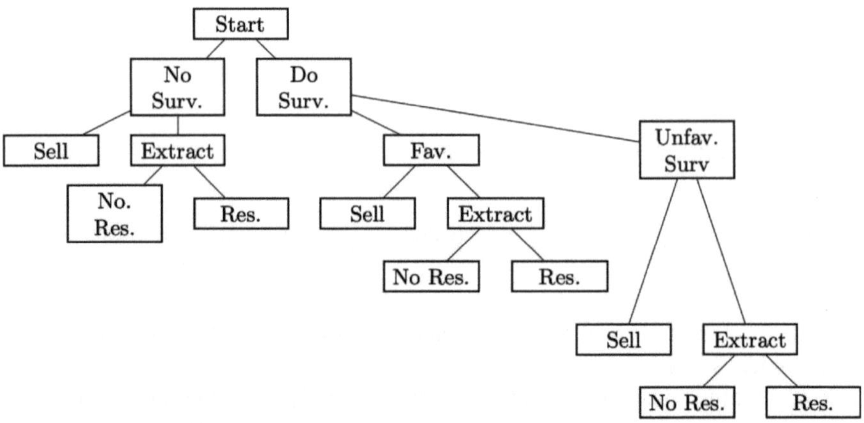

Fig. 12.2 Decision tree diagram

12.3 An Introduction to Multicriteria Optimization

Multicriteria optimization problems arise naturally when the goal is to find a satisfactory—that is, *efficient*—solution to programming problems that involve more than one objective function in the presence of fixed constraints.

Example 12.19. Consider the case of a new vehicle procurement program. For example, suppose that a company wants to make a purchasing decision between one of a collection of reasonably comparable vehicles.

The selection criteria at play include price, energy consumption, vehicle driving range, power and government subsidy. These factors are reflected in a collection of objective functions. The relevant selection constraints include constraints that reflect the total available budget for the procurement program, the average energy consumption of the entire fleet and the average vehicle driving range.

Here, it is clear that the companies aims are in conflict. For example, the company may like to minimize the cost and energy consumption while maximizing the vehicle driving range, available government subsidy and vehicle power.

Indeed, generally speaking maximizing the vehicle power would increase the energy consumption. Similarly, increasing driving range would likely not increase the vehicle power and would likely not decrease the energy consumption. Further, simultaneously trying to maximize driving range and power while minimizing energy consumption likely would see an increase and not decrease in the vehicle cost. Finally, the extent to which a government subsidy may be possible might increase as energy consumption decreases. But it might also involve other auxiliary factors as well.

The theory of *multicriteria mathematical programming* seeks to provide a mathematical framework to guide multicriteria decisions which arise in situations such as Example 12.19. A key role is played by the theory of *multiobjective linear programming problems.*

In more specific terms, a *canonical form multiobjective maximization linear programming problem* is formulated as

- Maximize the linear objective functions

$$f_1(x_1, \ldots, x_n), \ldots, f_k(x_1, \ldots, x_n)$$

- Subject to the linear constraints

$$\begin{array}{ccccccc} a_{11}x_1 & + & \ldots & + & a_{1n}x_n & \leq & b_1 \\ \vdots & & \vdots & & \vdots & & \vdots \\ a_{m1}x_1 & + & \ldots & + & a_{mn}x_n & \leq & b_m \end{array}$$

  and

$$x_1, \ldots, x_n \geq 0.$$

Similarly, a *canonical form multiobjective minimization linear programming problem* is formulated as

- Minimize the linear objective functions

$$g_1(x_1, \ldots, x_n), \ldots, g_k(x_1, \ldots, x_n)$$

- Subject to the linear constraints

$$\begin{array}{ccccccc} a_{11}x_1 & + & \ldots & + & a_{1n}x_n & \geq & b_1 \\ \vdots & & \vdots & & \vdots & & \vdots \\ a_{m1}x_1 & + & \ldots & + & a_{mn}x_n & \geq & b_m \end{array}$$

and

$$x_1 \geq 0, \ldots, x_n \geq 0.$$

As in the case of linear programming problems, the constraints in a multiobjective programming problem specify the feasible region. A typical running assumption in multiobjective programming problems is that no feasible solution simultaneously optimizes all objective functions (since otherwise such feasible solutions evidently give rise to optimal solutions). Since optimal solutions to multiobjective programming problems are in general not possible, instead we seek to find *efficient solutions*.

In more precise terms, the *standard lexicographic partial ordering* on $\mathbb{R}^k$ is defined by

$$\mathbf{x} = (x_1, \ldots, x_k) \leq_{lex} \mathbf{y} = (y_1, \ldots, y_k)$$

when $\mathbf{x} = \mathbf{y}$ or when the first nonzero coordinate of $\mathbf{x} - \mathbf{y}$ is negative. When no confusion is likely in what follows we denote $\leq_{lex}$ simply by $\leq$.

Now, given a multicriteria maximization linear programming problem as above, let

$$\mathbf{f} = (f_1(x_1, \ldots, x_n), \ldots, f_k(x_1, \ldots, x_n)).$$

Then fixing the standard lexicographic ordering on $\mathbb{R}^k$, a feasible solution $\hat{\mathbf{x}}$ is *maximally efficient* if

$$\mathbf{f}(\hat{\mathbf{x}}) \geq \mathbf{f}(\mathbf{x}) \text{ for all feasible solutions } \mathbf{x}.$$

Similarly, given a multicriteria minimization linear programming problem as above, let

$$\mathbf{g} = (g_1(x_1, \ldots, x_n), \ldots, g_k(x_1, \ldots, x_n)).$$

Then with respect to the standard lexicographic ordering on $\mathbb{R}^k$, a feasible solution $\hat{\mathbf{x}}$ is *minimally efficient* if

$$\mathbf{g}(\hat{\mathbf{x}}) \geq \mathbf{g}(\mathbf{x}) \text{ for all feasible solutions } \mathbf{x}.$$

In fixing the lexicographic ordering on $\mathbb{R}^k$ as above, in essence this gives a slight preference to the criteria that is specified by the given *ordered collection* of decision functions.

While the general study of such multicriteria linear optimization problems can be quite challenging there is a rich Simplex Method based approach to find all efficient solutions. As in the case of linear programming problems, the idea is to study the corner points of the feasible region and then to decide which give rise to efficient solutions. This is illustrated in Example 12.20.

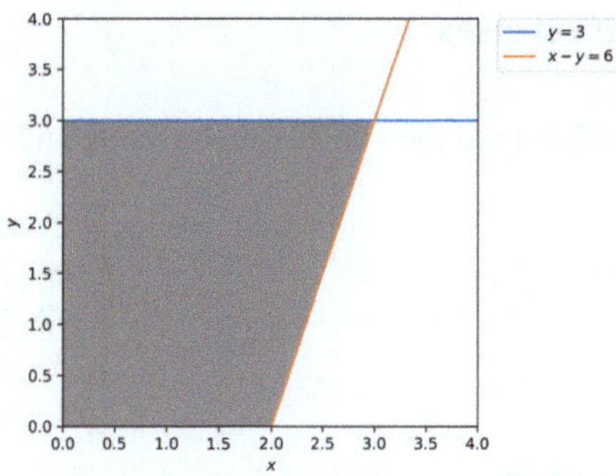

Fig. 12.3 Feasible region for Example 12.20

Table 12.4 Extreme points and objective function values for Example 12.20

Extreme point	Value of **f**
(0, 0)	(0, 0)
(0, 3)	(3, −6)
(2, 0)	(6, −2)
(2.5, 3)	(7.5, −7.5)

Example 12.20. Consider the canonical form multiobjective minimization linear programming problem

- Minimize the linear objective functions

$$3x + y \text{ and } -x - 2y$$

- Subject to the constraints

$$\begin{aligned} y &\le 3 \\ 3x - y &\le 6 \end{aligned}$$

  and

$$x, y \ge 0.$$

The feasible region is described in Figure 12.3.

It has extreme points (0, 0), (0, 3), (2, 0), (2.5, 3). To find minimally efficient solutions set

$$\mathbf{f}(x, y) = (3x + y, -x - 2y).$$

Then consider the following table of values (see Table 12.4).

The conclusion is that the extreme point (0, 3) is minimally efficient.

The Multicriteria Simplex Method ultimately refines the approach of Example 12.20. As output, it yields all efficient corner point solutions. One approach to this Multicriteria Simplex Method is described in the text [8]. A complementary exposition is described in [9].

Exercises

12.1 Find the von Neumann value and optimal mixed strategy for the matrix game

$$\begin{bmatrix} 2 & 1 & 4 & 2 \\ 1 & 2 & 1 & 1 \\ -2 & 6 & 3 & -2 \\ 3 & -3 & 5 & 1 \\ 1 & 2 & 2 & 1 \end{bmatrix}.$$

12.2 Suppose that two manufacturers currently are competing for sales in two different yet equally profitable product lines. In both cases, suppose that the sales volume for Manufacturer 2 is three times as large as that for Manufacturer 1. Both manufacturers intend to make improvements to their products; they need to decide on a marketing strategy.

Suppose that if both product improvements are developed simultaneously, then either manufacturer can have the two products ready for sale in 12 months. Another option is to develop only one product first in an attempt to get it marketed ahead of the competition. By doing this, Manufacturer 2 could have one product ready for sale in 9 months, whereas Manufacturer 1 would require 10 months as it has previous commitments for its production facilities. For either manufacturer, the second product could then be ready for sale in an additional 9 months.

For either product line, if both manufacturers market their improved models simultaneously, it is estimated that Manufacturer 1 would increase its share of the total future sales of this product by 8% of the total (from 25% to 33%). Similarly, Manufacturer 1 would increase its share by 20, 30 and 40% of the total if it marketed the product sooner than Manufacturer 2 by 2, 6 and 8 months, respectively. On the other hand, Manufacturer 1 would lose 4, 10, 12 and 14% of the total if Manufacturer 2 marketed it sooner by 1, 3, 7 and 10 months, respectively.

Formulate the above problem as a matrix game and then determine optimal mixed strategies for the respective manufacturers.

12.3 Suppose that a grocery store manager needs to replenish the strawberry supply. Suppose that the regular supplier can provide as many cases as needed. However, because these strawberries already are very ripe, they will need to be sold very soon and then discard any that remain unsold.

Suppose that the manager estimates that it will be possible to sell 12, 13, 14 or 15 cases in the very near future. Suppose that strawberries can be purchased for \$7 per case and then sold for \$18 per case. The manager now needs to decide how many cases to purchase.

Based on past consumer behaviour, the manager estimates that the prior probabilities are 0.1, 0.3, 0.4 and 0.2 for being able to sell 12, 13, 14 and 15 cases of strawberries in the very near future.

Table 12.5 Payoff table for Exercise 12.4

	Improving economy	Stable Economy	Worsening Economy
Strategy 1	\$30 million	\$5 million	−\$10 million
Strategy 2	\$40 million	\$10 million	−\$30 million
Strategy 3	−\$10 million	0	−\$15 million
Prior probability	0.1	0.5	0.4

(i) Develop a decision analysis formulation of this problem by identifying the decision alternatives, the states of nature and the payoff table.
(ii) How many cases of strawberries should be purchased if using the maximin payoff criterion?
(iii) How many cases should be purchased according to the maximum likelihood criterion?
(iv) How many cases should be purchased according to Bayes' decision rule?

12.4 Suppose that an investment firm believes in three possible scenarios over the span of a particular investing time period. These are: an improving economy, a stable economy and a worsening economy.

To invest based on these scenarios, the firm must choose between three different types of investment strategies. The first type is intended to perform very well in an improving economy and then only encounter a small loss in a worsening economy. The second type is intended to perform well in an improving economy but then perform poorly in a worsening economy. The third is intended to lose some money in an improving economy but then perform well in a worsening economy.

Suppose that the estimated profits and summarized in the payoff table, with prior probabilities, which is given in Table 12.5.

Which investment strategy should be taken under each of the following three criteria?

(i) Maximin payoff criterion.
(ii) Maximum likelihood criterion.
(iii) Bayes' decision rule.

Chapter 13
Introduction to Non-directed Networks

Overview

- Selected introduction to non-directed networks.
- Concepts of connectivity for non-directed networks.
- Introduction to trees and minimum spanning tree problem.

13.1 Non-directed Networks and Abstract Graphs

There is a rich abstract theory of *non-directed networks*. Here, and elsewhere, we interchangeably refer to non-directed networks as *abstract graphs with non-directed edges*.

We want to discuss some introductory parts of the general theory. We provide almost no proofs for the results that we state. Instead we refer to [15] or [11] for details. Many results within the context of undirected graphs admit analogs and/or extensions to the case of directed graphs. Here, and elsewhere we also interchangeably refer to directed networks as directed graphs.

Similar to the case of directed networks, abstract (non-directed) graphs arise in many practical problems. They thus oftentimes serve as models for real-world situations. A more formal definition of abstract non-directed graph is given in Definition 13.2. Example 13.1 below gives a representative motivational example.

Example 13.1 (Acquaintance graphs)**.** Define the *acquaintance graph* to be the graph where each vertex represents a person and where an edge connects two vertices if the corresponding people know each other.

The concept of acquaintance graph is flexible. For example, acquaintance graphs can be used to model people living in particular neighbourhoods, particular office spaces, particular countries or particular neighbourhoods that are connected by public transit hubs.

N. Grieve, *An Introduction to Mathematical Programming and Network Science*,
Springer Undergraduate Texts in Mathematics and Technology,
https://doi.org/10.1007/978-3-032-13330-4_13

In Definition 13.2, we formulate a number of abstract concepts which are useful in studying structural features of abstract graphs.

Definition 13.2.

- An undirected graph G consists of a *vertex set* $V(G)$, an *edge set* $E(G)$ and a relation that associates two vertices (not necessarily distinct) to each edge.
- We *draw* a graph by representing vertices by points and then representing edges as non-directed line segments and/or loops between the corresponding vertices.
- An *isomorphism between two graphs* G and H is a bijection f that maps $V(G)$ to $V(H)$, $E(G)$ to $E(H)$ and which is such that each edge of G with endpoints $u, v \in V(G)$ is mapped to an edge of H with endpoints $f(u)$ and $f(v)$. If G and H are isomorphic, then we also write $G \simeq H$.

Adjacency matrices for abstract graphs are defined in the following way.

Definition 13.3. Let G be an undirected graph with vertices $v_1, \ldots, v_n$. The *adjacency matrix* of G is the $n \times n$ matrix A whose (i, j) entry is

$$[A]_{ij} = \begin{cases} 1 & \text{if } v_i \text{ and } v_j \text{ are adjacent} \\ 0 & \text{otherwise.} \end{cases}$$

In terms of determining the extent to which two graphs G_1 and G_2, with respective adjacency matrices A_1 and A_2, are isomorphic, this is a difficult problem in general. One characterization is that G_1 and G_2 are isomorphic if and only if there exists a permutation matrix P which is such that

$$PA_1P^{-1} = A_2.$$

These concepts are illustrated in the following examples.

Example 13.4. The graph G with vertices $v_1, \ldots, v_8$ and adjacency matrix

$$\begin{bmatrix} 0\,1\,1\,0\,0\,0\,1\,0 \\ 1\,0\,0\,1\,0\,0\,0\,1 \\ 1\,0\,0\,1\,1\,0\,0\,0 \\ 0\,1\,1\,0\,0\,1\,0\,0 \\ 0\,0\,1\,0\,0\,1\,1\,0 \\ 0\,0\,1\,0\,0\,1\,1\,0 \\ 0\,0\,0\,1\,1\,0\,0\,1 \\ 1\,0\,0\,0\,1\,0\,0\,1 \\ 0\,1\,0\,0\,0\,1\,1\,0 \end{bmatrix}$$

can be pictured as in Figure 13.1.

It turns out to be isomorphic to the graph H with vertices $w_1, \ldots, w_8$, adjacency matrix

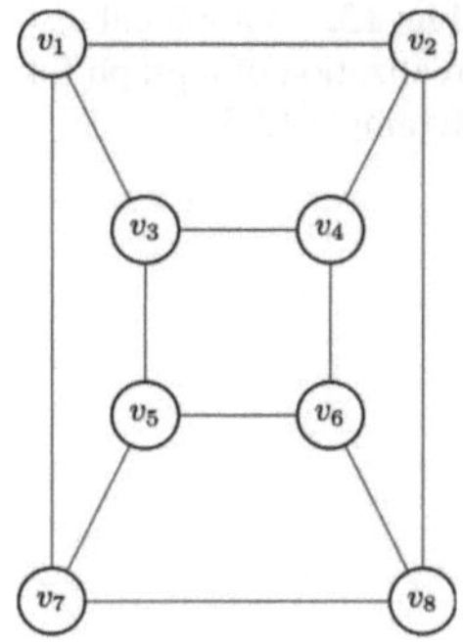

Fig. 13.1 Graphical realization of a graph for Example 13.4

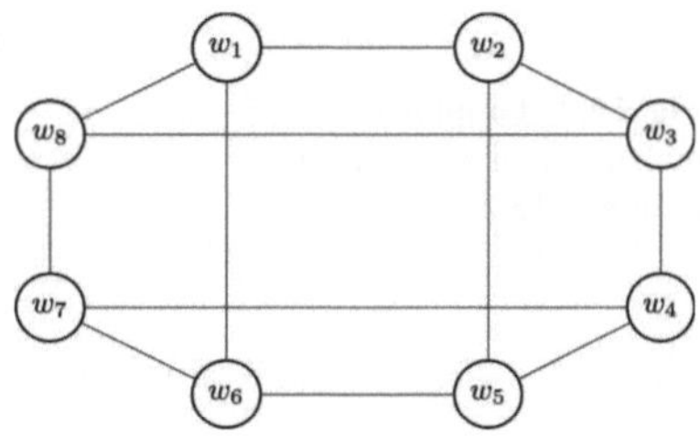

Fig. 13.2 Graphical realization of a graph for Example 13.4

$$\begin{bmatrix} 0&1&0&0&0&1&0&1\\ 1&0&1&0&1&0&0&0\\ 0&1&0&1&0&0&0&1\\ 0&0&1&0&1&0&1&0\\ 0&1&0&1&0&1&0&0\\ 1&0&0&0&1&0&1&0\\ 0&0&0&1&0&1&0&1\\ 1&0&1&0&0&0&1&0 \end{bmatrix}$$

and which can be pictured as in Figure 13.2.

Example 13.5. Consider the graph G_1 which is pictured in Figure 13.3. It has adjacency matrix

$$\begin{bmatrix} 0&1&0&1&0&1&0&0&0&0\\ 1&0&1&0&0&0&0&0&1&0\\ 0&1&0&0&1&0&1&0&0&0\\ 1&0&0&0&0&0&1&1&0&0\\ 0&0&1&0&0&1&0&1&0&0\\ 1&0&0&0&1&0&0&0&0&1\\ 0&0&1&1&0&0&0&0&0&1\\ 0&0&0&1&1&0&0&1&0&0\\ 0&1&0&0&0&0&0&1&0&1\\ 0&0&0&0&0&1&1&0&1&0 \end{bmatrix}.$$

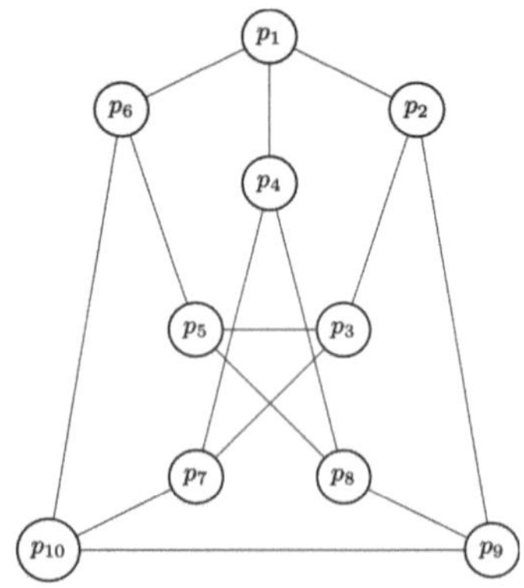

Fig. 13.3 Graphical realization of a graph for Example 13.5

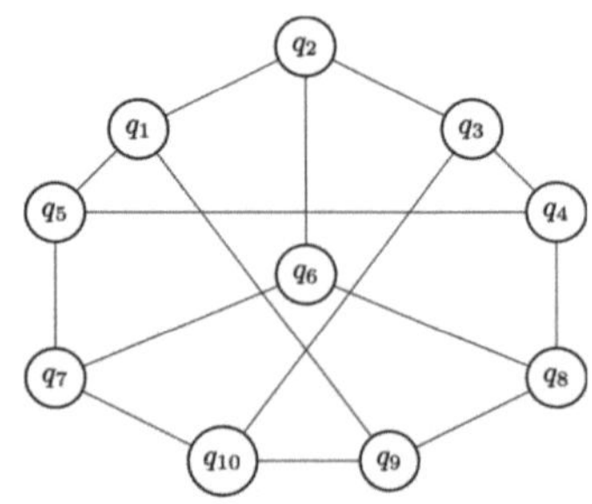

Fig. 13.4 Graphical realization of a graph for Example 13.5

Consider also the graph G_2 which is described in Figure 13.4.

It has adjacency matrix

$$\begin{bmatrix} 0&1&0&0&1&0&0&0&1&0\\ 1&0&1&0&0&1&0&0&0&0\\ 0&1&0&1&0&0&0&0&0&1\\ 0&0&1&0&1&0&0&1&0&0\\ 1&0&0&1&0&0&1&0&0&0\\ 0&1&0&0&0&0&1&1&0&0\\ 0&0&0&0&1&1&0&0&0&1\\ 0&0&0&1&0&1&0&0&1&0\\ 1&0&0&0&0&0&0&1&0&1\\ 0&0&1&0&0&0&1&0&1&0 \end{bmatrix}.$$

Finally, consider the graph G_3 which is described in Figure 13.5.

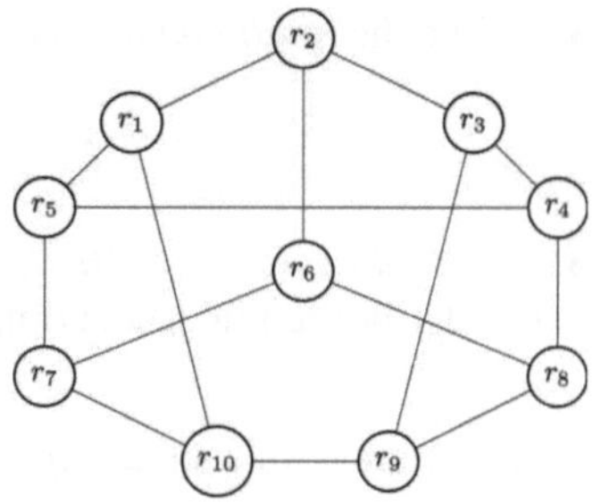

Fig. 13.5 Graphical realization of a graph for Example 13.5

It has adjacency matrix

$$\begin{bmatrix} 0\,1\,0\,0\,1\,0\,0\,0\,0\,1 \\ 1\,0\,1\,0\,0\,1\,0\,0\,0\,0 \\ 0\,1\,0\,1\,0\,0\,0\,0\,1\,0 \\ 0\,0\,1\,0\,1\,0\,0\,0\,1\,0 \\ 1\,0\,0\,1\,0\,0\,1\,0\,0\,0 \\ 0\,1\,0\,0\,0\,0\,1\,1\,0\,0 \\ 0\,0\,0\,0\,1\,1\,0\,0\,0\,1 \\ 0\,0\,0\,1\,0\,1\,0\,0\,1\,0 \\ 0\,0\,1\,0\,0\,0\,0\,1\,0\,1 \\ 1\,0\,0\,0\,0\,0\,1\,0\,1\,0 \end{bmatrix}.$$

It then turns out that

$$G_1 \simeq G_2 \not\simeq G_3.$$

We now introduce some of the most basic vocabulary for concepts that are attached to edges and vertices of abstract graphs.

Definition 13.6. Let G be a graph with vertex set $V(G)$ and edge set $E(G)$. Since G is assumed to be undirected, if $u, v \in V(G)$ are endpoints of some edge then we may write either uv or vu to denote the edge that they determine.

- Two vertices are called *adjacent* if they are joined by an edge. If a vertex is an endpoint of an edge, then we say that this edge is *incident* to this given vertex.
- The *neighbourhood* of $v \in V(G)$ is defined to be

$$N(v) := \{u \in V : vu \in E(G)\}.$$

- The *degree* of a vertex $v \in V(G)$ is defined to be

$$\deg(v) := \#\{\text{edges } e \in E(G) \text{ that are incident with } v\}.$$

- Define the *minimum degree* of a graph G to be

$$\delta(G) := \min\{\deg(v) : v \in V(G)\}.$$

- If $n := \#V(G)$, then the *degree sequence* is defined to be the n-term sequence, usually written in descending order, of the vertex degrees.

We are now in a situation to state the first fundamental theorem of graph theory.

Theorem 13.7. *In a graph G, the sum of the degrees of the vertices is equal to twice the number of edges. Consequently, the number of vertices of odd degree is even.*

The idea behind the proof of Theorem 13.7 is as follows. Let

$$S := \sum_{v \in V} \deg(v).$$

In order to determine S, we count each edge exactly twice. Therefore

$$S = 2|E|.$$

Now, since S is even, it follows that the number of vertices of odd degree must be even. (We need an even sum of odd numbers to ensure that the sum of all odd numbers is an even number.)

13.2 Concepts of Connectivity in Graphs

In this section, we introduce some of the most basic notions that allow for concepts of *connectivity* in non-directed graphs.

First of all, a *walk* in a graph G is a sequence of (not necessarily distinct) vertices $v_1, \ldots, v_k$ such that $v_i v_{i+1} \in E(G)$ for $i = 1, \ldots, k-1$. Second, if the vertices in a walk are distinct, then the walk is called a *path*. A *cycle* is a path $v_1, \ldots, v_k$ together with an additional edge $v_k v_1$.

The *distance* $d(u, v)$ between two vertices $u, v \in V(G)$ in a graph G is defined to be the number of edges in a shortest $u - v$ path in G.

The *diameter* of G is

$$\operatorname{diam}(G) := \max_{u,v \in V(G)} d(u, v).$$

The *eccentricity* of a vertex $v \in V(G)$ is

$$\operatorname{ecc}(v) := \max_{u \in V(G)} d(u, v).$$

The *radius* of G, is defined to be

$$\operatorname{rad}(G) := \min_{v \in V(G)} \operatorname{ecc}(v).$$

The starting point to the study of connectivity of a graph is the following theorem.

Theorem 13.8. *In a graph G, if $u, v \in V(G)$, then each $u - v$ walk contains a $u - v$ path.*

It turns out that powers of the adjacency matrix facilitate calculation of distance and enumeration of walks in graphs.

Theorem 13.9. *Let G be a graph with vertices labelled $v_1, \ldots, v_n$. Let A be its adjacency matrix. For each positive integer k, the (i, j) entry of A^k is equal to the number of walks from v_i to v_j that use exactly k edges.*

Corollary 13.10. *Let G be a graph with labelled vertices $v_1, \ldots, v_n$. Let A be its adjacency matrix. If $d(v_i, v_j) = \ell$, then $[A^k]_{i,j} = 0$ for $1 \leq k < \ell$.*

It turns out that adjacency matrices can be used to calculate eccentricity of vertices, radius and diameter of graphs.

Theorem 13.11. *Let G be a connected graph with vertices $v_1, \ldots, v_n$ and adjacency matrix A. For each positive integer k, define*

$$S_k := I + A + A^2 + \cdots + A^k.$$

- *If k is the smallest positive integer such that row j of S_k contains no zeros, then* $\operatorname{ecc}(v_j) = k$.
- *If r is the smallest positive integer such that all entries of at least one row of S_r are positive, then* $\operatorname{rad}(G) = r$.
- *If m is the smallest positive integer such that all entries S_m are positive, then* $\operatorname{diam}(G) = m$.

13.3 Degree Sequences for Undirected and Directed Graphs

As interesting invariant of undirected, respectively, directed graphs, are their *degree sequences*, respectively, list of *degree pairs*. We define what is meant by such concepts here, in what follows.

Definition 13.12. The *degree sequence* of an undirected graph is the list of vertex degrees, usually written in non-increasing order $d_1 \geq \cdots \geq d_n$.

Each undirected graph has a degree sequence $d_1 \geq \cdots \geq d_n$; but conversely given a collection of such nonnegative integers $d_1 \geq \cdots \geq d_n$ is there a graph with this degree sequence? This question is answered in the following theorem.

Theorem 13.13. *A collection of nonnegative integers $d_1 \geq \cdots \geq d_n$ are the vertex degrees of some undirected graph if and only if $\sum d_i$ is even.*

We now wish to discuss the topic of degree sequences for directed graphs. First, we give an abstract definition of directed graph. In fact, we have already encountered directed graphs in the form of directed networks.

Definition 13.14. A *directed graph* or *digraph* G consists of a *vertex set* $V(G)$ an *edge set* $E(G)$ and a function that assigns a (directed) edge to each ordered pair of vertices (u, v).

The vertex u is called the *head* of the directed edge (u, v); the vertex v is called the *tail*. The *out-degree* $d^+(v)$ is the number of edges with tail v. The *in-degree* $d^-(v)$ is the number of edges with head v.

The key observation in the proof of the following theorem is the fact that each edge has exactly one tail and exactly one head.

Theorem 13.15. *In a digraph G,*

$$\sum_{v \in V(G)} d^+(v) = |E(G)| = \sum_{v \in V(G)} d^-(v).$$

There is also a concept of degree sequences for digraphs. This is achieved via the notion of *degree pairs*.

Definition 13.16. For a digraph, we consider its list of *degree pairs* $(d^+(v_i), d^-(v_i))$.

When does a list of such pairs arise from some digraph? An answer is provided in Theorem 13.17 below.

Theorem 13.17. *A list of pairs of nonnegative integers is realizable as the degree pairs of a digraph if and only if the sum of the first coordinates equals the sum of the second coordinates.*

13.4 Selected Introduction to Trees

In this section, we introduce the concept of *trees*. Tress, interesting objects of their own right, can also be seen as the atomic pieces of more general graphs.

Definition 13.18. A *tree* is a *connected graph* that contains no *cycle*. The *leaves* of a tree are those vertices which have degree equal to one.

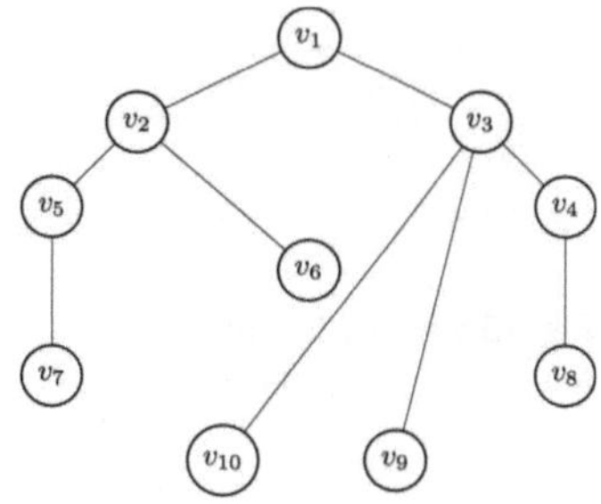

Fig. 13.6 Graphical realization of a graph for Example 13.19

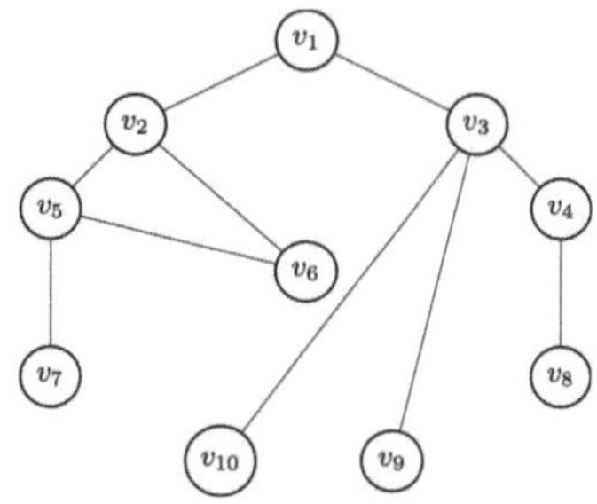

Fig. 13.7 Graphical realization of a graph for Example 13.19

Example 13.19. The graph given in Figure 13.6 is connected and contains no cycle. Hence it is a tree.

The graph given in Figure 13.7 contains a cycle $(v_2, v_5), (v_5, v_6), (v_6, v_2)$ and hence is not a tree.

In many ways, as mentioned, trees are the basic building blocks of abstract graphs quite generally. Trees themselves are also of an independent interest and can be used to model many real-world problems of interest.

There are several different abstract characterizations of trees.

Theorem 13.20. *Let G be an order n graph. The following assertions are equivalent.*

(i) G is a tree.
(ii) G is connected and contains $n - 1$ edges.
(iii) G is acyclic and contains $n - 1$ edges.

Counting trees

In order to count trees in a sensible way, it is important to rigidify the problem. *Cayley's Tree Formula* gives an enumeration of the number of distinct *labelled trees* of order n. Here, by a *labelled tree* of order n we mean a tree with a fixed vertex set $\{v_1, \ldots, v_n\}$.

Having defined the concept of labelled tree, Cayley's Tree Formula can be formulated in the following way.

Theorem 13.21 (Cayley's Tree Formula). *There are n^{n-2} distinct labelled trees of order n.*

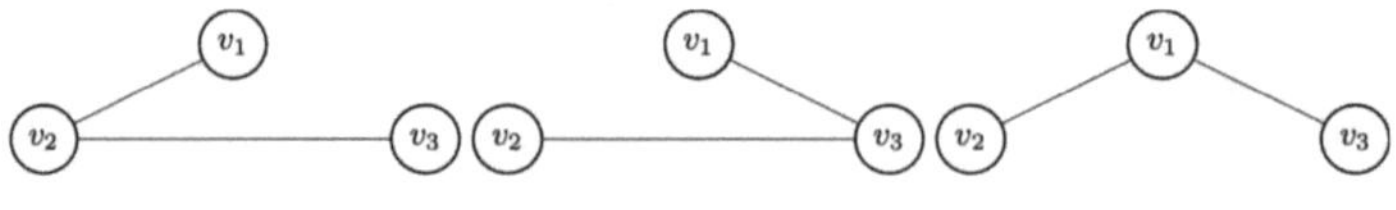

Fig. 13.8 Labelled trees on three vertices for Example 13.22

Fig. 13.9 Labelled trees on four vertices for Example 13.22

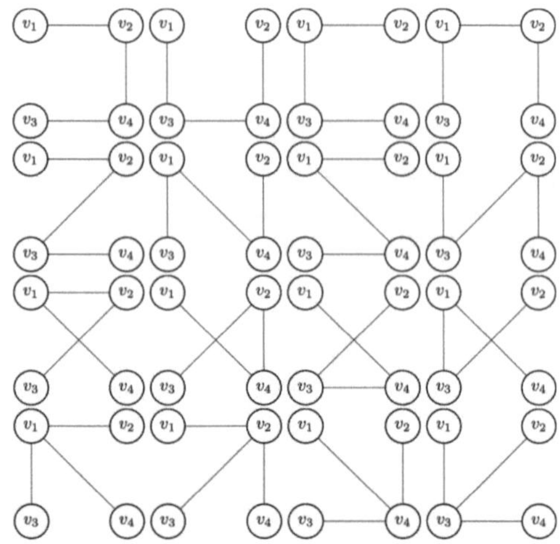

We want to discuss the idea of *Prüfer sequence* and outline its role in the proof of Cayley's Theorem about counting labelled trees. However, before doing so, let us illustrate the conclusion of Theorem 13.21 by way of Example 13.22.

Example 13.22. There are three distinct labelled trees on three vertices. They are described in Figure 13.8.

There are sixteen distinct labelled trees on four vertices. They are depicted in Figure 13.9.

Outline to prove Cayley's tree theorem via Prüfer sequences

Here, we outline the approach to proving Cayley's Tree Formula (Theorem 13.21) via Prüfer's method. There are two steps. These are Prüfer's method which assigns a suitable sequence to each labelled tree and then conversely to assign a labelled tree to each given Prüfer sequence. The key point is to show that these two methods are mutual inverses of each other.

To this end, the underlying idea is to put the set of labelled trees into bijection with the set of all length $n-2$ sequences with entries from the set $\{1, \dots, n\}$. There are n^{n-2} such sequences in total. This bijection is set up via the concept of *Prüfer sequence* assigned to a labelled graph (and conversely).

Prüfer's method starts with a labelled tree and as output produces its associated Prüfer sequence. The method is described in the following way.

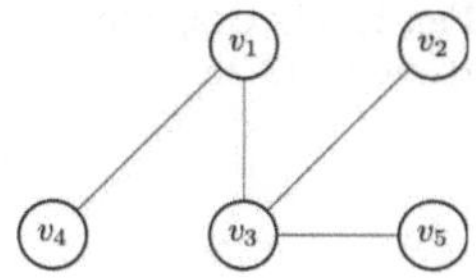

Fig. 13.10 Labelled tree for Example 13.23

Prüfer's method

Given a tree T with vertices labeled $1, \ldots, n$:

(i) Let $i = 0$ and let $T_0 = T$.
(ii) Find the leaf on T_i with the smallest label and call it v.
(iii) Record in the sequence the label of v's neighbour.
(iv) Remove v from T_i to create a new tree T_{i+1}.
(v) It $T_{i+1} = K_2$, the complete graph on 2-vertices, then stop. Otherwise increment i by 1 and go back to Step (ii).

Conversely, to assign each labelled tree to a Prüfer sequence proceed in the following way.

Assign a labelled tree a Prüfer sequence

Given a sequence $\sigma = a_1, \ldots, a_k$ of entries from the set $S := \{1, \ldots, k+2\}$:

(i) Draw $k+2$ vertices; label them $v_1, \ldots, v_{k+2}$.
(ii) Let $i = 0$, let $\sigma_0 = \sigma$ and let $S_0 = S$.
(iii) Let j be the smallest number in S_i that does not appear in the sequence σ_i.
(iv) Place an edge between vertex v_j and the vertex whose subscript appears first in the sequence σ_i.
(v) Remove the first number in the sequence σ_i to create a new sequence σ_{i+1}. Remove the element j from the set S_i to create a new set S_{i+1}.
(vi) If the sequence σ_{i+1} is empty, then place an edge between the two vertices whose subscripts are in S_{i+1} and then stop. Otherwise increment i by 1 and then return to Step (iii).

Example 13.23. The Prüfer sequence that is associated to the labelled tree that is described in Figure 13.10 is 3, 4, 1.

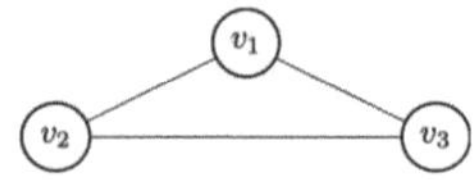

Fig. 13.11 Graph for Example 13.25

13.5 Introduction to the Matrix Tree Theorem

In this section, we describe the *Matrix Tree Theorem.* It gives a tool for counting the number of *spanning trees* in a given labelled connected graph.

First of all, a *spanning tree* in a connected graph G is a tree T in G that contains each of G's vertices. In order for state the Matrix Tree Theorem (see Theorem 13.24), we first need to introduce the concept of *degree matrix* for G. To this end, if $v_1, \dots, v_n$ are the vertices of G, then the *degree matrix* is the $n \times n$ matrix D that is defined by the condition that its (i, j) entry d_{ij} is

$$d_{ij} = \begin{cases} \deg(v_i) & \text{if } i = j \\ 0 & \text{otherwise.} \end{cases}$$

Let A be the adjacency matrix of G. Recall that A is the $n \times n$ matrix whose (i, j) entry is described as

$$a_{ij} = \begin{cases} 1 & \text{if } v_i \text{ and } v_j \text{ are adjacent.} \\ 0 & \text{otherwise.} \end{cases}$$

Finally, recall that given an $n \times n$ matrix M, its (i, j)*-cofactor* is defined to be

$$C_{ij} := (-1)^{i+j} \det(M(i\,|\,j)).$$

Here, $\det(M(i\,|\,j))$ represents the determinant of the $(n-1) \times (n-1)$ matrix that is obtained from M by deleting row i and column j.

With these preliminary concepts at hand, the *Matrix Tree Theorem* is formulated in the following way.

Theorem 13.24 (Matrix Tree Theorem). *Let G be a connected undirected graph with adjacency matrix A and degree matrix D. Then the number of unique spanning trees of G is equal to the value of any and every cofactor of the matrix $D - A$.*

The conclusion of Theorem 13.24 is illustrated in Example 13.25 below.

Example 13.25. The graph G is pictured in Figure 13.11 has three unique spanning trees.

They are described in Figure 13.12.

From the vantage point of the Matrix Tree Theorem (Theorem 13.24), the adjacency matrix A and degree matrix D are, respectively, described as

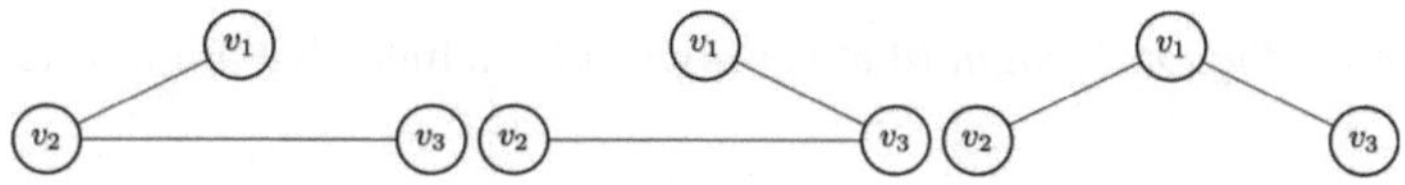

Fig. 13.12 Spanning trees for Example 13.25

$$A = \begin{bmatrix} 0 & 1 & 1 \\ 1 & 0 & 1 \\ 1 & 1 & 0 \end{bmatrix}$$

and

$$D = \begin{bmatrix} 2 & 0 & 0 \\ 0 & 2 & 0 \\ 0 & 0 & 2 \end{bmatrix}.$$

The matrix $A - D$ is described as

$$A - D = \begin{bmatrix} -2 & 1 & 1 \\ 1 & -2 & 1 \\ 1 & 1 & -2 \end{bmatrix}$$

and each of its cofactors are indeed equal to 3.

13.6 Introduction to Minimum Spanning Tree Problem

Another type of network and/or graph theory type problem to be aware of is the *minimum spanning tree problem*. The setting is that of an undirected and connected network together with some *measure of positive length* (e.g. distance, cost, time) that is associated to each *link* (that is *edge*). For the *minimum spanning tree problem* the idea is that the chosen links must provide a path between *each pair of nodes*; the idea is to do this in the most cost effective way.

13.6.1 Formulation of the Minimum Spanning Tree Problem

The *minimal weighted spanning tree problem*, or simply the *minimum spanning tree problem*, is formulated in the following way.

Definition 13.26 (Minimum spanning tree problem)**.** Given the nodes (that is vertices) of a network, the *minimum spanning tree problem* seeks to design the links in the network by inserting enough links to satisfy the requirement that there will be a path between each pair of nodes. Further, satisfy these requirements in a way that *minimizes the total length* (that is *weight*) of the links that are inserted into the network.

In Section 8.9, we have already encountered a form of *minimal spanning tree* in a graph. This was done via the *shortest path network problem.*

It turns out that there are various approaches to formulating and solving the minimal spanning tree problem. We describe one such approach in Section 13.6.3. First, in Section 13.6.2, we discuss selected applications.

13.6.2 Brief Discussion About Selected Applications of Minimum Spanning Tree Problem and Network Problems

While some applications of the *shortest path problem* include

- minimize the total distance travelled between two given nodes; and/or
- minimize the total time of a sequence of activities,

in many ways, by contrast, the *minimum spanning tree problem* is more general in nature.

Moreover, some important applications of minimum spanning tree problem include

- design of telecommunication networks while minimizing the total cost of cables;
- design of transportation network while minimizing the total cost of the links;
- design of electrical power transmission network; and/or
- design of a network of pipelines to connect a number of different locations.

In order to translate network problems into abstract graph theoretic problems, it is helpful to introduce some specific network/graph specific terminology.

This is achieved in Definition 13.27 below. Specifically, we define what is means for a network to be *connected.* Further, we define precisely what is meant by *spanning tree* in a connected network.

Definition 13.27.

- Two nodes in an undirected network are called *connected* if the network contains at least one undirected path between them.
- A *connected network* is a network where every pair of nodes is connected.
- A *spanning tree* in a connected network with n nodes is a connected subnetwork that contains no undirected cycles.

Note that each such spanning tree will have exactly $n - 1$ arcs. Indeed, the reason is that this is the minimum number of arcs needed to have a connected network and the maximum number possible without having undirected cycles.

13.6.3 A Method for the Minimum Spanning Tree Problem

It turns out, that the minimum spanning tree problem can be solved easily via the following *greedy algorithm*. In some sense, this algorithm can be seen as a direct graph theoretic solution to an unconstrained linear optimization problem. This highlights the interactions between linear programming approaches to solve network optimization problems compared to graph theoretic techniques.

A method to solve the minimum weight spanning tree problem is summarized in the following way.

A method to solve the Minimum Weight Spanning Tree problem

(i) Select any node arbitrarily and then connect it with an edge to the nearest distinct node.

(ii) Find the unconnected node that is closest (that is has minimal weight) to a connected node and then connect these two nodes with an edge. Repeat this step until all nodes have been connected.

(iii) Ties for nearest distance in the above two steps may be broken in an arbitrary way.

Remark 13.28.

- Encountering ties at any point in the method may indicate the possibility that more than one optimal solution may exist.
- On small sized examples, it turns out that oftentimes the fastest way to execute the method is via a direct graphical approach.

Example 13.29. We consider the effect of running the minimum spanning tree method, starting with the vertex v_1, to the undirected weighted network that is given in Figure 13.13. The evolving minimum weight spanning tree is described in Figure 13.14. The minimum spanning weight is thus 19.

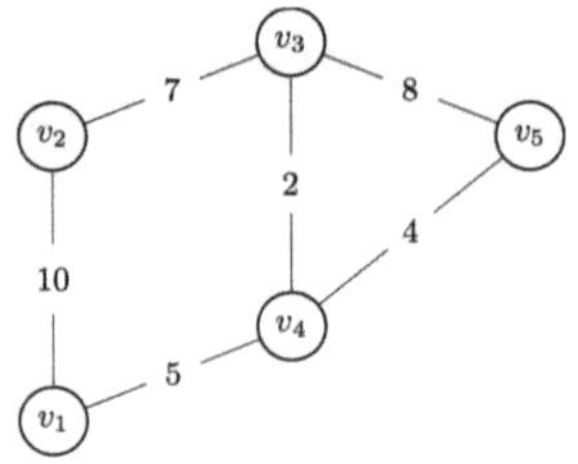

Fig. 13.13 Undirected weighted network for Example 13.29

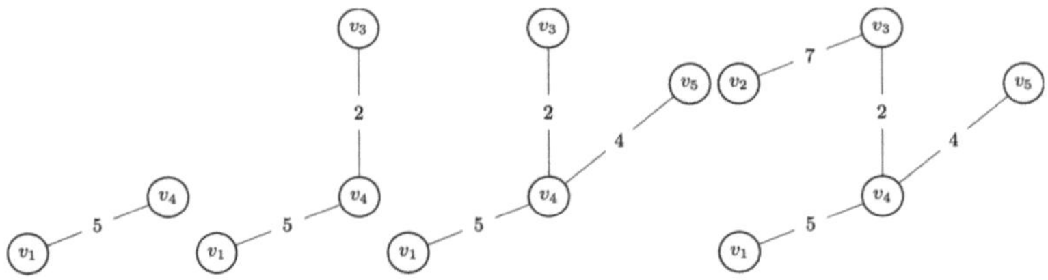

Fig. 13.14 Evolving minimum weight spanning tree for Example 13.29

Exercises

13.1 Ten people are seated around a circular table. Each person shakes hands with everyone at the table. Draw an (undirected) graph that models this situation.

13.2 If v is a vertex of an undirected graph G, then let $N(v)$ denote its set of neighbours and $\deg(v)$ its degree. Prove or disprove that the cardinality of $N(v)$ equals $\deg(v)$. What about if G is simple (that is if G has no loops nor multiple edges)?

13.3 Define the *intersection graph* of a collection of sets $A_1, \ldots, A_n$ to be the undirected graph that has a vertex for each of these sets and has an edge connecting the vertices representing two sets if these sets have a nonempty intersection. Suppose that $n = 5$ and that $A_1 = \{0, 2, 4, 6, 8\}$, $A_2 = \{0, 1, 2, 3, 4\}$, $A_3 = \{1, 3, 5, 7, 9\}$, $A_4 = \{5, 6, 7, 8, 9\}$ and $A_5 = \{0, 1, 8, 9\}$. Construct the corresponding intersection graph in this context.

13.4 The *degree sequence* of an undirected graph is the sequence of the degrees of the vertices in non-increasing order. A sequence $d_1, \ldots, d_n$ is called *graphic* if it is the degree sequence of a simple graph. Determine which of the following sequences is graphic. For those that are, draw a graph having the given degree sequence:

(i) $5, 4, 3, 2, 1, 0$;
(ii) $5, 4, 3, 2, 1$; and
(iii) $4, 4, 3, 2, 1$.

Table 13.1 Weighted vertex table for Exercise 13.6

edge	$\{a, b\}$	$\{a, c\}$	$\{c, f\}$	$\{f, g\}$	$\{g, e\}$	$\{e, b\}$	$\{a, d\}$	$\{b, d\}$	$\{f, d\}$	$\{f, d\}$	$\{d, g\}$	$\{c, d\}$	$\{d, e\}$
weight	7	2	5	7	6	4	4	4	3	6	5	3	2

13.5 Prove that every undirected graph has an even number of odd vertices.

13.6 Let G be the weighted graph with vertices a, b, c, d, e, f, g and weighted edges as indicated in Table 13.1. Give a graphical realization of G and find a minimum weight spanning tree.

Appendix A
Solving Systems of Linear Equations

Overview

- Consistent, inconsistent and equivalent linear systems of equations.
- Gaussian Elimination and row echelon form augmented matrices.
- Gauss-Jordan Elimination and reduced row echelon form augmented matrices.
- Solving linear systems of equations via Gauss-Jordan Elimination.
- Introduction to Matrix Algebra.

A.1 Introduction to $m \times n$ Systems of Linear Equations

Consider the case of an $m \times n$ *system of linear equations*

$$\begin{aligned} a_{11}x_1 + \ldots + a_{1n}x_n &= b_1 \\ \vdots \quad\quad \vdots \quad\quad \vdots \quad &\quad \vdots \\ a_{m1}x_1 + \ldots + a_{mn}x_n &= b_m \end{aligned}$$

where $a_{ij} \in \mathbb{R}$, for $i = 1, \ldots, m$, $j = 1, \ldots, n$ and $b_1, \ldots, b_m \in \mathbb{R}$.

Recall, that each such system has either

(i) a unique solution;
(ii) infinitely many solutions; or
(iii) no solutions.

In case that either (i) or (ii) hold true, then the system is said to be a *consistent system*. In case that (iii) holds true, then the system is said to be an *inconsistent system*.

N. Grieve, *An Introduction to Mathematical Programming and Network Science*, Springer Undergraduate Texts in Mathematics and Technology,
https://doi.org/10.1007/978-3-032-13330-4

Here by a *solution* to the given linear system of equations is meant values for the unknown variables $x_1, \ldots, x_n$ which satisfy each of the given m linear equations. The *solution set* of a system of linear equations is the set of all solutions of the system. By *solving the system of linear equations*, is meant the process of finding the solution set. By the *general solution* to a system of linear equations, we mean a description of all solutions.

Definition A.1. Two $m \times n$ linear systems of equations

$$\begin{array}{ccccc} a_{11}x_1 & + \ldots + & a_{1n}x_n & = & b_1 \\ \vdots & \vdots & \vdots & & \vdots \\ a_{m1}x_1 & + \ldots + & a_{mn}x_n & = & b_m \end{array}$$

and

$$\begin{array}{ccccc} a'_{11}x_1 & + \ldots + & a'_{1n}x_n & = & b'_1 \\ \vdots & \vdots & \vdots & & \vdots \\ a'_{m1}x_1 & + \ldots + & a'_{mn}x_n & = & b'_m \end{array}$$

are said to be *equivalent* if they have the same solution set.

In solving a given $m \times n$ linear system of equations

$$\begin{array}{ccccc} a_{11}x_1 & + \ldots + & a_{1n}x_n & = & b_1 \\ \vdots & \vdots & \vdots & & \vdots \\ a_{m1}x_1 & + \ldots + & a_{mn}x_n & = & b_m \end{array}$$

the starting point for the method that we employ here is to consider the corresponding *augmented matrix*

$$[A|\mathbf{b}] = \left[\begin{array}{ccc|c} a_{11} & \ldots & a_{1n} & b_1 \\ \vdots & & \vdots & \vdots \\ a_{m1} & \ldots & a_{mn} & b_m \end{array}\right].$$

Conversely, we note that each such augmented matrix $[A|\mathbf{b}]$ determines uniquely an $m \times n$ linear system of equations.

In solving $m \times n$ linear systems of equations, the idea is to use a sequence of row operations to convert the given $m \times n$ system of equations into an equivalent system of linear equations but which has corresponding augmented matrix in *reduced row echelon form*. The concept of *reduced row echelon form augmented matrix* is defined in Section A.3. Here, we define the concept of *elementary row operations* applied to a given augmented matrix.

Definition A.2. By an *elementary row operation* applied to a given augmented matrix $[A|\mathbf{b}]$ is meant one of the following operations:

(i) interchange two rows ($R_i \leftrightarrow R_j$);
(ii) multiply a row by a nonzero constant ($R_i \rightarrow kR_i$); or
(iii) add a multiple of a row to another row ($R_i \rightarrow R_i + kR_j$).

The main theoretical justification for solving $m \times n$ systems of linear equations via the method of Gauss-Jordan Elimination is provided by the following fundamental result.

Theorem A.3. *The augmented matrices of equivalent $m \times n$ systems of linear equations can be transformed from one to another via some finite sequence of elementary row operations.*

A.2 A Brief Overview of Gaussian Elimination

Here, we describe the method of *Gaussian Elimination* which is a key practical tool for determining a *row echelon form* of a given *augmented matrix*

$$[A|\mathbf{b}] = \left[\begin{array}{ccc|c} a_{11} & \dots & a_{1n} & b_1 \\ \vdots & & \vdots & \vdots \\ a_{m1} & \dots & a_{mn} & b_m \end{array}\right]$$

that corresponds to a given $m \times n$ linear system of equations

$$\begin{array}{ccccc} a_{11}x_1 & + \dots + & a_{1n}x_n & = & b_1 \\ \vdots & \vdots & \vdots & & \vdots \\ a_{m1}x_1 & + \dots + & a_{mn}x_n & = & b_m. \end{array}$$

Definition A.4. Let $[A|\mathbf{b}]$ be the augmented matrix of an $m \times n$ linear system of equations.

- An entry of the augmented matrix $[A|\mathbf{b}]$ is a *leading entry* or a *pivot entry* if it is the leftmost nonzero entry of a row.
- Each column of $[A|\mathbf{b}]$ that contains a pivot entry is called a *pivot column.*
- The augmented matrix $[A|\mathbf{b}]$ is said to be in *row echelon form* if the following two conditions hold true.

  (i) All rows of zeros are below all nonzero rows.
  (ii) Each leading entry of a row is in a column that is to the right of the leading entry of any row that is above it.

In Section A.3, we apply the method of Gaussian Elimination in our formulation of Gauss-Jordan Elimination. In turn, these two methods give rise to practical procedures for solving $m \times n$ systems of linear equations.

The method of Gaussian Elimination applied to a given augmented matrix $[A|\mathbf{b}]$ produces as output an augmented row echelon form matrix. It can be formulated in the following way.

Gaussian Elimination applied to $[A|\mathbf{b}]$

Let $[A|\mathbf{b}]$ be the augmented matrix of a given $m \times n$ linear system of equations

$$\begin{array}{ccccccc} a_{11}x_1 & + & \dots & + & a_{1n}x_n & = & b_1 \\ \vdots & & \vdots & & \vdots & & \vdots \\ a_{m1}x_1 & + & \dots & + & a_{mn}x_n & = & b_m \end{array}$$

Assume that this system is not trivial—that is at least one $a_{ij} \neq 0$.

(i) Starting from the left, find the first nonzero column. The position at the top of this column will be the first pivot entry. Switch rows if needed so as to place a non-zero number in this first pivot position.
(ii) Use row operations to make the entries below the first pivot entry (in the first pivot column) equal to zero.
(iii) Ignoring the row containing the first pivot entry, repeat Steps (i) and (ii) with the remaining rows. Repeat the process until there are no more non-zero rows left.

A.3 A Brief Overview of Gauss-Jordan Elimination

The method of Gauss-Jordan Elimination is a key practical tool for determining a *row echelon form* of a given *augmented coefficient matrix*

$$[A|\mathbf{b}] = \left[\begin{array}{ccc|c} a_{11} & \dots & a_{1n} & b_1 \\ \vdots & & \vdots & \vdots \\ a_{m1} & \dots & a_{mn} & b_m \end{array}\right]$$

that corresponds to a given $m \times n$ linear system of equations

$$\begin{array}{ccccccc} a_{11}x_1 & + & \ldots & + & a_{1n}x_n & = & b_1 \\ \vdots & & \vdots & & \vdots & & \vdots \\ a_{m1}x_1 & + & \ldots & + & a_{mn}x_n & = & b_m. \end{array}$$

The concept of *reduced row echelon form augmented matrix* is made precise as follows.

Definition A.5. The augmented matrix $[A|\mathbf{b}]$ is said to be in *reduced row echelon form* if the following three conditions hold true.

(i) it is in row echelon form;
(ii) each leading entry is equal to 1; and
(iii) all entries that are above a leading entry are equal to zero.

The method of Gauss-Jordan Elimination applied to a given augmented matrix $[A|\mathbf{b}]$ produces as output an augmented reduced row echelon form matrix. It can be formulated as follows.

Gauss-Jordan Elimination to solve $[A|\mathbf{b}]$

Let $[A|\mathbf{b}]$ be the augmented matrix of a given $m \times n$ linear system of equations

$$\begin{array}{ccccccc} a_{11}x_1 & + & \ldots & + & a_{1n}x_n & = & b_1 \\ \vdots & & \vdots & & \vdots & & \vdots \\ a_{m1}x_1 & + & \ldots & + & a_{mn}x_n & = & b_m \end{array}$$

Assume that this system is not trivial—that is at least one $a_{ij} \neq 0$.

(i) First use Gaussian Elimination to determine a row echelon form augmented matrix

$$\mathrm{REF}([A|\mathbf{b}]) = [A'|\mathbf{b}'].$$

(ii) Moving from right to left in the row echelon form augmented matrix $\mathrm{REF}([A|\mathbf{b}])$ consider each leading entry. For each such leading entry, without changing the row that contains it or any row below it, use row operations to create zeros in the column above the leading entry. Finally, divide the row by its pivot entry to make the pivot entry equal to 1. The end result is the reduced row echelon form, denoted by $\mathrm{RREF}([A|\mathbf{b}])$, of the augmented matrix $[A|\mathbf{b}]$.

A.4 Solving Linear Systems of Equations via Gauss-Jordan Elimination

The method of *Gauss-Jordan Elimination* applied to a given $m \times n$ linear system of equations

$$\begin{array}{ccccc} a_{11}x_1 & + \ldots + & a_{1n}x_n & = & b_1 \\ \vdots & \vdots & \vdots & & \vdots \\ a_{m1}x_1 & + \ldots + & a_{mn}x_n & = & b_m \end{array}$$

(should any solutions exist) is described in the following way.

Gauss-Jordan Elimination to solve $A\mathbf{x} = \mathbf{b}$

(i) First determine the corresponding reduced row echelon form augmented matrix RREF($[A|\mathbf{b}]$).
(ii) The linear system of equations that corresponds to RREF($[A|\mathbf{b}]$) is either evidently consistent or not; should it be consistent then it is readily solved, regarding the variables that correspond to non-pivot columns as free variables and using back substitution as required.

In terms of structural results for solutions to consistent linear equations, the main result is described in Theorem A.6 below.

Theorem A.6. *Consider the case of a consistent $m \times n$ linear system of equations*

$$\begin{array}{ccccc} a_{11}x_1 & + \ldots + & a_{1n}x_n & = & b_1 \\ \vdots & \vdots & \vdots & & \vdots \\ a_{m1}x_1 & + \ldots + & a_{mn}x_n & = & b_m. \end{array}$$

Then the general solution

$$\mathbf{x} = \begin{bmatrix} x_1 \\ \vdots \\ x_n \end{bmatrix} \in \mathbb{R}^n$$

can be described as

$$\mathbf{x} = \mathbf{x}_h + \mathbf{x}_p \in \mathbb{R}^n$$

where

$$\mathbf{x}_h = \begin{bmatrix} x_{h1} \\ \vdots \\ x_{hn} \end{bmatrix} \in \mathbb{R}^n$$

is the general solution to the associated homogeneous system of equations

$$\begin{array}{ccccc} a_{11}x_1 & + \dots + & a_{1n}x_n & = 0 \\ \vdots & \vdots & \vdots & \vdots \\ a_{m1}x_1 & + \dots + & a_{mn}x_n & = 0 \end{array}$$

and where

$$\mathbf{x}_p = \begin{bmatrix} x_{p1} \\ \vdots \\ x_{pn} \end{bmatrix} \in \mathbb{R}^n$$

is any solution to the given linear system of equations.

A.5 Selected Examples

Example A.7. Solve the linear system of equations

$$\begin{aligned} x + 2y - z &= 3 \\ 2x + 3y + z &= 1. \end{aligned}$$

using Gaussian Elimination. The augmented matrix of the system is

$$\left[\begin{array}{ccc|c} 1 & 2 & -1 & 3 \\ 2 & 3 & 1 & 1 \end{array}\right] \xrightarrow{R_2 \to R_2 - 2R_1} \left[\begin{array}{ccc|c} 1 & 2 & -1 & 3 \\ 0 & -1 & 3 & -5 \end{array}\right].$$

Setting $z = t$ for $t \in \mathbb{R}$ gives

$$y = 3t + 5, x = -2(3t + 5) + t + 3 = -5t - 7.$$

A parametric description of the solution set is thus given by

$$x = -5t - 7, y = 3t + 5 \text{ and } z = t \text{ for } t \in \mathbb{R}.$$

Example A.8. The 2×4 homogeneous linear system of equations

$$\begin{aligned} 2x_1 + x_2 + x_3 + 4x_4 &= 0 \\ x_1 + 2x_2 - x_3 + 5x_4 &= 0. \end{aligned}$$

has the augmented matrix

$$[A|\mathbf{b}] = \left[\begin{array}{cccc|c} 2 & 1 & 1 & 4 & 0 \\ 1 & 2 & -1 & 5 & 0 \end{array}\right].$$

Sequentially performing the following series of elementary row operations:

- $R_1 \leftrightarrow R_2$;
- $R_2 \rightarrow R_2 - 2R_1$;
- $R_2 \rightarrow -\frac{1}{3}R_2$;
- $R_1 \rightarrow R_1 - 2R_2$

yields the RREF augmented matrix

$$\left[\begin{array}{cccc|c} 1 & 0 & 1 & 1 & 0 \\ 0 & 1 & -1 & 2 & 0 \end{array}\right].$$

The general solution to this homogeneous linear system of equations can be described parametrically as

$$x_1 = -x_3 - x_4;\ x_2 = x_3 - 2x_4;\ x_3 = s,\ x_4 = t \text{ for } t, s \in \mathbb{R}.$$

Example A.9. The 2×3 linear system of equations

$$\begin{aligned} x_1 + 4x_2 + 3x_3 &= 2 \\ 3x_1 + 12x_2 + 9x_3 &= 6 \end{aligned}$$

has a particular solution given by

$$\mathbf{x}_p = \begin{bmatrix} x_{p1} \\ x_{p2} \\ x_{p3} \end{bmatrix} = \begin{bmatrix} 2 \\ 0 \\ 0 \end{bmatrix}.$$

Moreover, the general solution of the associated homogeneous linear system of equations is described by

$$\mathbf{x}_h = \begin{bmatrix} x_{h1} \\ x_{h2} \\ x_{h3} \end{bmatrix} = \begin{bmatrix} -4s - 3t \\ s \\ t \end{bmatrix} \text{ for } s, t \in \mathbb{R}.$$

Indeed, that $\mathbf{x}_p$ is a particular solution to the given system of linear equations can be checked directly. To see that $\mathbf{x}_h$ indeed describes the general solution consider the coefficient matrix of the associated homogeneous system of linear equations

$$A = \begin{bmatrix} 1 & 4 & 3 \\ 3 & 12 & 9 \end{bmatrix}.$$

It has RREF

$$\text{RREF}(A) = \begin{bmatrix} 1 & 4 & 3 \\ 0 & 0 & 9 \end{bmatrix}.$$

It is thus clear that the general solution to the associated homogeneous system of equations is described by $\mathbf{x}_h$ as desired.

Now, given $\mathbf{x}_p$ and $\mathbf{x}_h$, it follows that the general solution of the given 2×3 system of linear equations is described by

$$\mathbf{x} = \mathbf{x}_p + \mathbf{x}_h = \begin{bmatrix} 2 - 4s - 3t \\ s \\ t \end{bmatrix} \quad \text{for } s, t \in \mathbb{R}.$$

Indeed, we simply check that

$$2 - 4s - 3t + 4s + 4t = 2$$

and

$$3(2 - 4s - 3t) + 12s + 9t = 6.$$

A.6 Matrix Operations

An $m \times n$ *matrix*, with real entries, is a rectangular array of numbers

$$A = \begin{bmatrix} a_{11} & \dots & a_{1n} \\ \vdots & & \vdots \\ a_{m1} & \dots & a_{mn} \end{bmatrix} \in \mathbb{R}^{m \times n}.$$

The real numbers $a_{ij} \in \mathbb{R}$ are called the *entries* of A. If $n = m$, then A is called a *square matrix*.

Example A.10. The $n \times n$ *identity matrix* $I_{n \times n}$ is the $n \times n$ matrix that has 1's on the diagonal and 0's elsewhere. Given two matrices

$$A = [a_{ij}] \in \mathbb{R}^{m \times n} \text{ and } B = [b_{ij}] \in \mathbb{R}^{r \times s}$$

we write

$$A = B$$

if and only if

$$m = r, n = s \text{ and } a_{ij} = b_{ij} \text{ for all } i, j.$$

The operations of matrix *scalar multiplication*, *matrix addition* and *matrix subtraction* are defined in the following way.

Given an $m \times n$ matrix

$$A = [a_{ij}] \in \mathbb{R}^{m \times n}$$

and a real number $c \in \mathbb{R}$, the matrix obtained from A by *scalar multiplication* by c, is defined to be the $m \times n$ matrix

$$c \cdot A = [ca_{ij}] \in \mathbb{R}^{m \times n}.$$

To define addition of matrices, given two $m \times n$ matrices

$$A = [a_{ij}],\ B = [b_{ij}] \in \mathbb{R}^{m \times n},$$

their sum is the $m \times n$ matrix

$$A + B := [a_{ij} + b_{ij}] \in \mathbb{R}^{m \times n}.$$

Especially, the matrix $A + B$ has i, j entry $a_{ij} + b_{ij}$.

To define subtraction of matrices, the matrix $A - B$ is defined to be

$$A - B := A + (-1) \cdot B \in \mathbb{R}^{m \times n}.$$

(It is not possible to add and/or subtract matrices of incompatible sizes.)

Example A.11. If

$$A = \begin{bmatrix} 1 & 4 & 0 \\ -2 & 6 & 5 \end{bmatrix}$$

then

$$2A = \begin{bmatrix} 2 & 8 & 0 \\ -4 & 12 & 10 \end{bmatrix}.$$

Example A.12. If

$$A = \begin{bmatrix} 1 & 4 & 0 \\ -2 & 6 & 5 \end{bmatrix},\ B = \begin{bmatrix} -3 & 1 & -1 \\ 3 & 0 & 2 \end{bmatrix} \text{ and } C = \begin{bmatrix} 4 & 3 \\ 2 & 1 \end{bmatrix}$$

then

$$A + B = \begin{bmatrix} -2 & 5 & -1 \\ 1 & 6 & 7 \end{bmatrix}$$

whereas $B + C$ is not defined. Similarly,

$$A - B := A + (-B) = \begin{bmatrix} 4 & 3 & 1 \\ -5 & 6 & 3 \end{bmatrix}.$$

In order to define *matrix multiplication*, let

$$A = [a_{ij}] \in \mathbb{R}^{m \times n}$$

be an $m \times n$ matrix and let

$$B = [b_{ij}] \in \mathbb{R}^{n \times r}$$

be an $n \times r$ matrix. Then the matrix

$$C := AB = [c_{ij}] \in \mathbb{R}^{m \times r}$$

is defined to be the $m \times r$ matrix that has i, j entry

$$c_{ij} = a_{i1}b_{1j} + \cdots + a_{in}b_{nj}.$$

Thus,

$$c_{ij} = \sum_{k=1}^{n} a_{ik}b_{kj}.$$

Example A.13. If

$$A = \begin{bmatrix} 1 & 3 & -1 \\ -2 & -1 & 1 \end{bmatrix} \text{ and } B = \begin{bmatrix} -4 & 0 & 3 & -1 \\ 5 & -2 & -1 & 1 \\ -1 & 2 & 0 & 6 \end{bmatrix},$$

then

$$AB = \begin{bmatrix} 12 & -8 & 0 & -4 \\ 2 & 4 & -5 & 7 \end{bmatrix}.$$

Example A.14. The linear system of equations

$$x_1 - 2x_2 + 3x_3 = 5$$

$$-x_1 + 3x_2 + x_3 = 1$$

$$2x_1 - x_2 + 4x_3 = 14$$

can be expressed in matrix terms as

$$\begin{bmatrix} 1 & -2 & 3 \\ -1 & 3 & 1 \\ 2 & -1 & 4 \end{bmatrix} \begin{bmatrix} x_1 \\ x_2 \\ x_3 \end{bmatrix} = \begin{bmatrix} 5 \\ 1 \\ 14 \end{bmatrix}.$$

More generally, each $m \times n$ system of equations

$$\begin{array}{ccccccc} a_{11}x_1 & + & \dots & + & a_{1n}x_n & = & b_1 \\ \vdots & & \vdots & & \vdots & & \vdots \\ a_{m1}x_1 & + & \dots & + & a_{mn}x_n & = & b_m \end{array}$$

can be expressed in matrix terms as

$$A\mathbf{x} = \mathbf{b}$$

where

$$A = \begin{bmatrix} a_{11} & \dots & a_{1n} \\ \vdots & \vdots & \vdots \\ a_{m1} & \dots & a_{mn} \end{bmatrix}, \mathbf{x} = \begin{bmatrix} x_1 \\ \vdots \\ x_n \end{bmatrix} \text{ and } \mathbf{b} = \begin{bmatrix} b_1 \\ \vdots \\ b_n \end{bmatrix}.$$

Aside from its frequent use in applications, matrix multiplication is a surprisingly complicated operation that is of an independent interest.

The operation of *matrix transpose* is another important operation that arises in practice. Especially, if

$$A = [a_{ij}] \in \mathbb{R}^{m \times n}$$

is an $m \times n$ matrix, then its *transpose* A^T is the $n \times m$ matrix

$$A^T = [a_{ji}] \in \mathbb{R}^{n \times m}.$$

Example A.15. If

$$A = \begin{bmatrix} 1 & 3 & 2 \\ 5 & 0 & 1 \end{bmatrix},$$

then

$$A^T = \begin{bmatrix} 1 & 5 \\ 3 & 0 \\ 2 & 1 \end{bmatrix}.$$

Example A.16. A square matrix

$$A = [a_{ij}] \in \mathbb{R}^{n \times n}$$

is *symmetric* if

$$A^T = A.$$

As a concrete example, the matrix

$$A = \begin{bmatrix} 1 & 3 & 2 \\ 3 & 5 & 0 \\ 2 & 0 & 4 \end{bmatrix}$$

is symmetric, since

$$A = A^T.$$

On the other hand, if

$$B = \begin{bmatrix} 1 & 2 \\ -1 & 3 \end{bmatrix}$$

then

$$B \neq B^T.$$

Thus, the matrix B is not symmetric.

Appendix B
Convexity and Optimization of Smooth Functions in Dimensions One and Two

Overview

- Local and global extreme values in dimension one.
- Local extrema in dimension two.
- Selected examples.

B.1 Local and Global Extreme Values in Dimension One

We describe, briefly in what follows, the main results that surround local and global extreme values for single variable functions.

B.1.1 Extreme Value Theorem and Closed Interval Method

Let $c \in \mathbb{R}$ be in the domain $D \subseteq \mathbb{R}$ of a function

$$f : D \to \mathbb{R}.$$

Recall, that $f(c)$ is the *absolute maximum* value of f on D if

$$f(c) \geq f(x) \text{ for all } x \in D;$$

$f(c)$ is an *absolute minimum* value of f on D if

$$f(c) \leq f(x) \text{ for all } x \in D.$$

N. Grieve, *An Introduction to Mathematical Programming and Network Science*, Springer Undergraduate Texts in Mathematics and Technology,
https://doi.org/10.1007/978-3-032-13330-4

On the other hand, $f(c)$ is a *local maximum value* of f if

$$f(c) \geq f(x) \text{ for all } x \text{ near } c;$$

whereas $f(c)$ is a local minimum value of f if

$$f(c) \leq f(x) \text{ for all } x \text{ near } c.$$

Theorem B.1. *If $f(x)$ is continuous on a closed interval $[a, b]$, then f attains an absolute maximum value $f(c)$ and an absolute minimum $f(d)$ at some $c, d \in [a, b]$.*

Recall, that the *critical numbers* of f are those $c \in D$ which are such that $f'(c) = 0$ or which are such that $f'(c)$ does not exist.

Theorem B.2. *If $f(x)$ has a local maximum or a local minimum at c, then c is a critical number of f.*

Theorem B.3 (Closed interval method). *To find the absolute maximum and absolute minimum values of a continuous function $f(x)$ on a closed interval $[a, b]$ proceed in the following way.*

(i) Find the values of $f(x)$ at all critical points $x = c \in (a, b)$.
(ii) Find the values of $f(x)$ at $x = a$ and $x = b$.
(iii) The largest values from Steps 1 and 2 is the absolute maximum value, whereas the smallest of these values is the absolute minimum value.

B.1.2 First and Second Derivative Tests for Local Extreme Values

In terms of finding local extreme values, the main results are the first and second derivative tests. Respectively, these results are formulated in the following way.

Theorem B.4 (First derivative test for $f(x)$). *Let*

$$f : D \to \mathbb{R}$$

be a continuous function with domain $D \subseteq \mathbb{R}$. Suppose that c is a critical number of f. The following assertions hold true.

(i) If $f'(x)$ changes from positive to negative at $x = c$, then $f(c)$ is a local maximum.
(ii) If $f'(x)$ changes from negative to positive at $x = c$, then $f(c)$ is a local minimum.
(iii) If $f'(x)$ does not change sign when passing from left to right at $x = c$, then $f(x)$ has no local maximum or local minimum at $x = c$.

Theorem B.5 (Second derivative test for $f(x)$). *Suppose that*

$$f: D \to \mathbb{R} \text{ for } D \subseteq \mathbb{R}$$

is a twice differentiable function and suppose that $f''(x)$ is continuous near $x = c$. Then the following assertions hold true.

(i) If $f'(c) = 0$ and $f''(c) > 0$, then $f(c)$ is a local minimum.
(ii) If $f'(c) = 0$ and $f''(c) < 0$, then $f(c)$ is a local maximum.

B.2 Local Extrema in Dimension Two

Here, we recall the second derivative test for the case of a two variable twice differentiable function

$$f(x, y): D \to \mathbb{R} \text{ for } D \subseteq \mathbb{R}^2.$$

First of all, we say that $f(x, y)$ has a *local maximum* at an interior point $(a, b) \in D$ if

$$f(x, y) \leq f(a, b) \text{ for all } (x, y) \text{ near } (a, b);$$

$f(x, y)$ has a *local minimum* and $f(a, b)$ is a *local minimum* at an interior point $(a, b) \in D$ if

$$f(x, y) \geq f(a, b) \text{ for all } (x, y) \text{ near } (a, b).$$

Further, if

$$f(a, b) \geq f(x, y) \text{ for all } (x, y) \in D,$$

then $f(a, b)$ is an *absolute maximum*; if

$$f(a, b) \leq f(x, y) \text{ for all } (x, y) \in D,$$

then $f(a, b)$ is an *absolute minimum*. Finally, recall that the *gradient vector* $\nabla f(x, y)$ is defined to be

$$\nabla f(x, y) := \langle f_x(x, y), f_y(x, y) \rangle.$$

With these preliminaries at hand, the second derivative test for $f(x, y)$ is formulated in the following way.

Theorem B.6 (Second derivative test for $f(x, y)$). *Let $f(x, y)$ be a twice differentiable function. If $f(x, y)$ has a local maximum or a local minimum value at $(x, y) = (a, b)$, then $\nabla f(a, b) = \mathbf{0}$; equivalently $f_x(a, b) = 0$ and $f_y(a, b) = 0$. Set*

$$D = D(a, b) := \begin{vmatrix} f_{xx}(a, b) & f_{xy}(a, b) \\ f_{yx}(a, b) & f_{yy}(a, b) \end{vmatrix} = f_{xx}(a, b) f_{yy}(a, b) - (f_{xy}(a, b))^2.$$

Then the following assertions hold true.

(i) If $D > 0$ *and* $f_{xx}(a, b) > 0$, *then* $f(a, b)$ *is a local minimum.*
(ii) If $D > 0$ *and* $f_{xx}(a, b) < 0$, *then* $f(a, b)$ *is a local maximum.*
(iii) If $D < 0$, *then* $f(a, b)$ *is not a local maximum or a local minimum. In this case,* (a, b) *is a* saddle point *and the graph of* $f(x, y)$ *crosses its tangent plane at* (a, b).

B.3 Selected Examples

Example B.7 below illustrates the single variable optimization results described in Sections B.1.1 and B.1.2.

Example B.7. Consider the function

$$y = f(x) = (x + 1)(x - 1)(x - 2)(x - 3) = x^4 - 5x^3 + 5x^2 + 5x - 6$$

which has graph described in Figure B.1.

Then

$$f'(x) = 4x^3 - 15x^2 + 10x + 5;$$

the critical points occur when

$$f'(x) = 0;$$

thus the critical points may be described as

$$x \approx -0.326345,\ 1.46909 \text{ and } 2.60725.$$

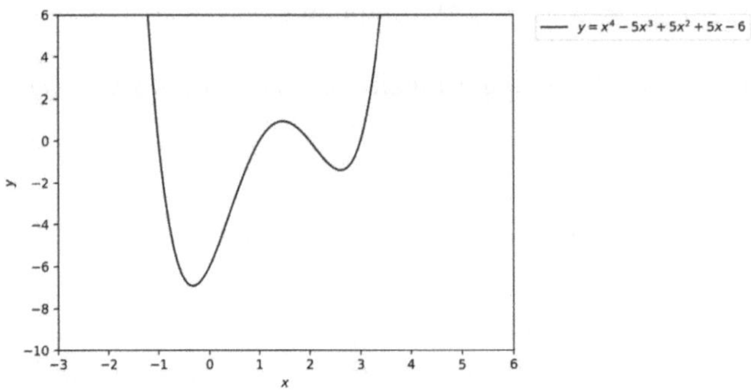

Fig. B.1 Graph for Example B.7

By applying the closed interval method on [0, 3], the conclusion is that $f(x)$ has maximum value

$$f(1.46909) \approx 0.941378$$

and minimum value

$$f(0) = -6.$$

On the other hand, considering $f(x)$ on $(-\infty, \infty)$, the conclusion upon inspecting the graph of $f(x)$ and as is confirmed by the first and/or second derivative tests, respectively, $f(x)$ has local minimum values

$$f(-0.326345) \approx -6.9141 \text{ and } f(2.60725) \approx -1.38275;$$

$f(x)$ has local maximum value

$$f(1.46909) \approx 0.9414.$$

Example B.8 below illustrates the two variable optimization results described in Section B.2.

Example B.8. Consider the function

$$z = f(x, y) = \frac{1}{4}x^4 + \frac{1}{4}y^4 - xy + 1$$

which has contour diagram described in Figure B.2. Then since

$$\nabla f(x, y) = \langle f_x, f_y \rangle = \langle x^3 - y, y^3 - x \rangle$$

there are three critical points

$$(0, 0), (1, 1) \text{ and } (-1, 1).$$

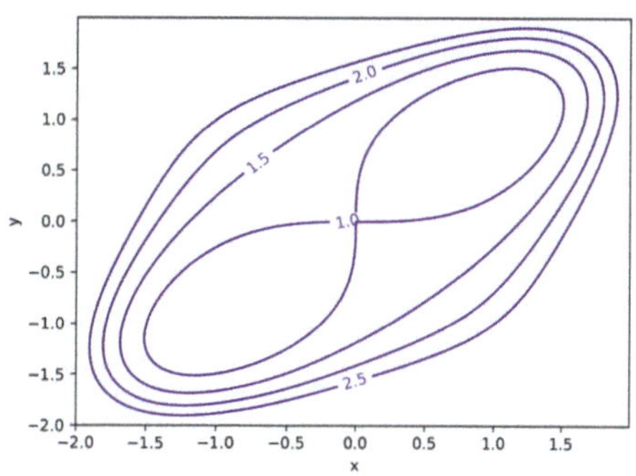

Fig. B.2 Graph for Example B.8

Further

$$f_{xx} = 3x^2,\ f_{yy} = 3y^2 \text{ and } f_{xy} = -1.$$

Thus

$$D(x, y) = f_{xx} f_{yy} - (f_{xy})^2 = 9x^2y^2 - 1.$$

As suggested by the contour diagram, (0, 0) is a saddle point, whereas $f(1, 1) = \frac{1}{2}$ and $f(-1, 1) = \frac{5}{2}$ are local minimums.

Indeed, since

$$D(0, 0) = -1 < 0,$$

$$D(1, 1) = 8 > 0$$

and

$$D(-1, 1) = 8 > 0,$$

this is confirmed as an application of the second derivative test.

Appendix C
Outline of Solutions to Selected Exercises

Overview

- Outline of solutions to selected exercises.

Chapter 1

1.1 The feasible region has five corner points $(9, 0)$, $(15/2, 9/2)$, $(3, 9)$, $(0, 15/2)$ and $(0, 0)$. The maximum value of the objective function is $Z(3, 9) = 210$.
1.2 The feasible region has three corner points $(0, 6)$, $(2, 4)$ and $(6, 2)$. The minimum value of the objective function is $Z(2, 4) = 110$.
1.3 The feasible region has five corner points $(0, 0)$, $(0, 800)$, $(800/3, 800)$, $(600, 300)$. The maximum value of the objective functions is $Z(600, 300) = 720{,}000$.
1.6 The feasible region has corner points $(0, 0)$, $(0, 1)$, $(1, 0)$ and $(2/3, 2/3)$. The maximum value is $Z(2/3, 2/3) = 10/3$.

Chapter 2

2.1 Let x_1 denote the amount of Type I meat and let x_2 denote the amount of Type II meat. The optimal solution occurs when $x_1 = .9375$, $x_2 = 0$, $Z(x_1, x_2) = 3$.
2.2 Denote the five items under consideration by the decision variables $x_1, \ldots, x_5$. The optimal solution is when $x_1 = 1$, $x_2 = 1$, $x_3 = 1$, $x_4 = 0$, $x_5 = 0$ in which case the objective function has value 230.
2.3 Setting x_i for $i = 1, \ldots, 6$ equal to the number of cashiers beginning work at the start of period i, the objective function has minimum value 45 which occurs when $x_1 = 2$, $x_2 = 18$, $x_3 = 0$, $x_4 = 20$, $x_5 = 0$ and $x_6 = 5$.
2.4 Introduce nonnegative decision variables $x_{11}, x_{12}, x_{13}, x_{21}, x_{22}, x_{23}$ to denote the amount of product shipped from Factory i to Retailer j for $i = 1, 2$ and $j = 1, 2, 3$. The cost function to be minimized is

$$C(x_{11}, \ldots, x_{23}) = 14x_{11} + 13x_{12} + 11x_{13} + 13x_{21} + 13x_{22} + 12x_{23}$$

N. Grieve, *An Introduction to Mathematical Programming and Network Science*, Springer Undergraduate Texts in Mathematics and Technology,
https://doi.org/10.1007/978-3-032-13330-4

subject to the constraints

$$x_{11} + x_{12} + x_{13} = 1200,\ x_{21} + x_{22} + x_{23} = 1000,\ x_{11} + x_{21} = 1000$$

$$x_{12} + x_{22} = 700 \text{ and } x_{13} + x_{23} = 500.$$

Chapter 3

3.1 Since $x, y, z, w \geq 0$, the inequality $5x + 4y + 3z + 2w \leq 1$ holds only for sufficiently small positive values of x, y, z, w. So, the constraint set is contained in some closed ball of sufficiently large radius. Thus the constraint set is bounded. Finding the extreme points amounts to solving each of the following five linear systems of equations:

(i) $5x + 4y + 3z + 2w = 1, x = 0, y = 0, z = 0$;
(ii) $5x + 4y + 3z + 2w = 1, x = 0, y = 0, w = 0$;
(iii) $5x + 4y + 3z + 2w = 1, x = 0, z = 0, w = 0$;
(iv) $5x + 4y + 3z + 2w = 1, y = 0, z = 0, w = 0$;
(v) $x = 0, y = 0, z = 0, w = 0$.

An optimal solution to the given linear programming problem is -2.

3.2 (i) The constraint set has corner points $(0, 0)$, $(2, 0)$, $(1, 2)$, $(2, 1)$ and $(0, 2)$. The maximum value of the objective function is $Z(1, 2) = 5$.

(ii) $(0, 0)$: $x_1 = 0, x_2 = 0$, $(2, 0)$: $x_1 = 2, x_2 = 0$, $(1, 2)$: $x_1 + x_2 = 3,\ x_2 = 2$, $(2, 1)$: $x_1 + x_2 = 3, x_1 = 2$, $(0, 2)$: $x_1 = 0, x_2 = 2$.

(iii) Solve each of the pairs of simultaneous equations.

(iv) $(0, 0, (0, 2)$ share $x_1 \geq 0$; $(0, 2), (1, 2)$ share $x_2 \leq 2$; $(1, 2), (2, 1)$ share $x_1 + x_2 \leq 3$, $(2, 1), (0, 2)$ share $x_1 \leq 2$, $(2, 0), (0, 0)$ share $x_2 \geq 0$.

3.3 The objective function is $Z = 1000x_1 + 2000x_2$. We calculate $Z(0, 0) = 0$, $Z(0, \frac{20}{8}) = 5000$, $Z(5, 5) = 15,000$, $Z(6, 4) = 14,000$ and $Z(8, 0) = 8,000$. Thus an optimal solution is $Z(5, 5) = 15,000$.

Chapter 4

4.1 (i) The corresponding slack linear programming problem can be expressed as

- Minimize the linear objective function

$$g(y_1, y_2, y_3) = 10y_1 + 6y_2 + 8y_3$$

- Subject to the constraints

$$y_1 + y_2 + 2y_3 = 16 + y_4$$

$$5y_1 + 3y_2 + 2y_3 = 21 + y_5$$

$$y_1, y_2, y_3, y_4, y_5 \geq 0.$$

(ii) The corresponding minimization tableau is

$$\begin{bmatrix} y_1 & 1 & 5 & 10 \\ y_2 & 1 & 3 & 6 \\ y_3 & 2 & 2 & 8 \\ & 16 & 21 & 0 \\ & y_4 & y_5 & \end{bmatrix}.$$

(iii) The corresponding negative transpose tableau is

$$\begin{bmatrix} y_1 & y_2 & y_3 & & \\ -1 & -1 & -2 & -16 & y_4 \\ -5 & -3 & -2 & -21 & y_5 \\ -10 & -6 & -8 & 0 & \end{bmatrix}.$$

4.3 (i) Set $g(y_1, y_2, y_3) = 20y_1 + 30y_2 + 25y_3$ and $f(x_1, x_2) = 200x_1 + 150x_2$. Then

$$g - f = 20y_1 + 30y_2 + 25y_3 - 200x_1 - 150x_2.$$

On the other hand,

$$s_1 = y_1 + 2y_2 + 2y_3 - 200, s_2 = 2y_1 + 2y_2 + y_3 - 150$$

and

$$t_1 = -x_1 - x_2 + 20, t_2 = -2x_1 - 2x_2 + 30 \text{ and } t_3 = -2x_1 - 2x_2 + 25.$$

Expanding and simplifying yields that

$$s_1x_1 + s_2x_2 + y_1t_1 + t_2t_2 + y_3t_3 = g - f$$

as desired.

(ii) Yes, they exhibit complementary slackness.

Chapter 5

5.3 Starting with the initial tableau

$$\begin{bmatrix} x_1 & x_2 & x_3 & & \\ 3 & 1 & 2 & 40 & x_4 \\ 2 & 1 & 2 & 40 & x_5 \\ 1 & 2 & 2 & 80 & x_6 \\ 2 & 4 & 3 & 0 & \end{bmatrix}$$

sequentially pivoting on the entries (2, 1), (3, 1) and (2, 1) yields a tableau from which can be extracted an optimal objective function value 160 with $x_1 = 0, x_2 = 40$ and $x_3 = 0$.
5.6 The negative transpose of the given minimization tableau is not a basic feasible tableau. Applying the Simplex Method yields a basic feasible tableau from which the optimal value of the given minimization programming problem is deduced to be 13500 which is achieved when $y_1 = 20$ and $y_2 = 15$.

Chapter 6
6.1 The given linear programming problem is unbounded.
6.2 The optimal value is $Z = -6$ and this is achieved when $x_1 = 6$, $x_2 = 0$ and $x_3 = 0$.
6.4 The optimal value is $Z = 3$ and this is achieved when $x_1 = 0, x_2 = 1$ and $x_3 = 0$.

Chapter 7
7.1 The maximum value of the objective function is $C = 360$ and this is achieved when $x_{13} = 10, x_{21} = 10, x_{22} = 10, x_{31} = 10, x_{33} = 20$ and all other $x_{ij} = 0$.
7.3 The maximum value of the objective function is $C = 185$ and this is achieved when $x_{11} = 5, x_{13} = 15, x_{21} = 15, x_{32} = 20, x_{43} = 15$ and all other $x_{ij} = 0$.
7.4 The maximum value of the objective function is $C = 19$ and this is achieved when $x_{16} = 1, x_{22} = 1, x_{34} = 1, x_{41} = 1, x_{53} = 1, x_{65} = 1$ and all other $x_{ij} = 0$.

Chapter 8
8.1 Introduce a decision variable x_{ij} for each edge in the network. Let v_s denote the source and v_d denote the destination. The shortest distance linear programming problem to be solved takes the form

- Minimize dist $= 4x_{01} + 2x_{03} + 3x_{12} + 2x_{14} + x_{23} + 3x_{24} + x_{31} + 4x_{34} + 3x_{45}$
- Subject to the constraints

$$x_{45} = 1$$

$$x_{01} + x_{31} - (x_{12} + x_{14}) = 0$$

$$x_{12} - (x_{23} + x_{24}) = 0$$

$$x_{03} + x_{23} - (x_{31} + x_{34}) = 0$$

$$x_{14} + x_{34} - x_{45} = 0$$

$$0 \le x_{01} \le 1, 0 \le x_{03} \le 1, 0 \le x_{12} \le 1, 0 \le x_{14} \le 1, 0 \le x_{23} \le 1,$$

$$0 \le x_{24} \le 1, 0 \le x_{31} \le 1, 0 \le x_{34} \le 1, 0 \le x_{45} \le 1$$

and all x_{ij} integers.

8.3 To convert the balanced transportation problem that is associated to the balanced transportation tableau into a minimal network flow problem, the idea is to augment the complete bi-partite (directed) graph that determined by the warehouses W_1, W_2, W_3 and the markets M_1, M_2, M_3, M_4 with a source vertex v_s and a drain vertex v_d.

The end result can be encoded the following weight adjacency matrix

	v_s	W_1	W_2	W_3	M_1	M_2	M_3	M_4	v_d
v_s	(40, 0)	(40, 0)	(100, 0)						
W_1					(40, 2)	(40, 1)	(40, 2)	(40, 0)	
W_2					(40, 2)	(40, 1)	(40, 2)	(40, 0)	
W_3					(100, 1)	(100, 2)	(100, 9)	(100, 0)	
M_1									(40, 0)
M_2									(50, 0)
M_3									(20, 0)
M_4									(70, 0)
v_d									

The balanced transportation to be solved is

- Minimize

$$Z = 2x_{12} + x_{21} + 2x_{23} + 2x_{21} + x_{22} + 2x_{23} + x_{31} + 2x_{32} + 9x_{33}$$

- Subject to the demand constraints

$$x_{11} + x_{21} + x_{31} = 40$$

$$x_{21} + x_{22} + x_{23} = 50$$

$$x_{31} + x_{32} + x_{33} = 20$$

$$x_{41} + x_{42} + x_{43} = 70$$

and subject to the warehouse constraints

$$x_{11} + x_{12} + x_{13}x_{14} = 40$$

$$x_{21} + x_{22} + x_{23} + x_{24} = 40$$

$$x_{31} + x_{32} + x_{33} + x_{34} = 100$$

and finally together with the nonnegativity constraints $x_{ij} \geq 0$.

Upon expanding out the minimal-cost-flow network problem that is determined by the above weight adjacency matrix, the end result is seen to be equivalent to the above balanced transportation problem.

8.4 The net flows are described as

$$\phi(v_1) = x_{61} - x_{12} - x_{15}$$

$$\phi(v_2) = x_{12} + x_{32} + x_{52} - x_{23}$$

$$\phi(v_3) = x_{23} - x_{32} - x_{34} - x_{35}$$

$$\phi(v_4) = x_{34} + x_{54} - x_{47}$$

$$\phi(v_5) = x_{15} + x_{35} + x_{75} - x_{54} - x_{52}$$

$$\phi(v_6) = x_{67} - x_{76} - x_{61}$$

$$\phi(v_7) = x_{47} + x_{67} - x_{76}.$$

Chapter 9

9.1 For the particular problem at hand

$$f(x) = 48x^5 + 42x^3 + 3.5x - 16x^6 - 61x^4 - 16.5x^2,$$

$x_1 = 1$ and $\epsilon = 0.001$. Several iterations of Newton's Method gives an optimal value for $f(x)$ at $x^* \approx 0.501$ with $f(x^*) \approx 0.313$. (Actually, a quicker convergence of Newton's Method can be obtained with more informed choice of start value x_1. For example, a start value of $x_1 = .6$ is appropriate.)

9.2 For the particular problem at hand,

$$f_{x_1} = \frac{1}{x_1},\ f_{x_2} = -2x_2,\ f_{x_1x_1} = -\frac{1}{x_1^2},\ f_{x_2x_2} = -2,\ f_{x_1x_2} = f_{x_2x_1} = 0.$$

Therefore

$$\text{Hess}(-f) = \begin{bmatrix} \frac{1}{x_1^2} & 0 \\ 0 & 2 \end{bmatrix}$$

and

$$|\text{Hess}(-f)| = \frac{2}{x_1^2} > 0,\ \frac{1}{x^2} > 0 \text{ and } 2 > 0$$

on some open convex subset of $\mathbb{R}^2$. Therefore $-f$ is strictly convex on some open convex subset of $\mathbb{R}^2$. Therefore f is concave.

Chapter 10

10.1 Setting $f_1(x_1, x_2) = x_1 - 10,\ f_2(x_1, x_2) = x_2 - 15,\ f_3 = -x_1,\ f_4 = -x_2$, the Lagrangian is given by

$$\begin{aligned} L(\mathbf{v}, \mathbf{x}) &= f(\mathbf{x}) + v_1 f_1(\mathbf{x}) + v_2 f_2(\mathbf{x}) + v_3 f_3(\mathbf{x}) + v_4 f_4(\mathbf{x}) \\ &= -12x_1 + x_1^2 - 50x_2 + x_2^2 + v_1(x_1 - 10) + v_2(x_2 - 15) - v_3 x_1 - v_4 x_2. \end{aligned}$$

10.2 Setting $f_1(x_1, x_2) = x_1 + 2x_2 - 3$, $f_2(x_1, x_2) = -x_1$ and $f_3(x_1, x_2) = -x_2$, the Lagrangian is given by

$$L(\mathbf{v}, \mathbf{x}) = f(\mathbf{x}) + v_1 f_1(\mathbf{x}) + v_2 f_2(\mathbf{x}) + v_3 f_3(\mathbf{x}) = \ln(x_1 + 1) - x_2^2 + v_1(x_1 + 2x_3 - 3) - v_2 x_1 - v_3 x_2.$$

10.4 See [14, §28, p. 279].

Chapter 11

11.2 A relevant contour diagram is given in Figure C.1.

11.3 A relevant contour diagram superimposed over the feasible region is given in Figure C.2.

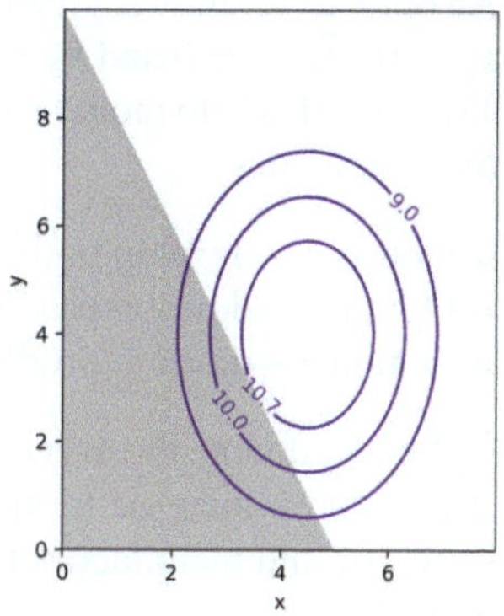

Fig. C.1 Figure for solution to Exercise 11.2

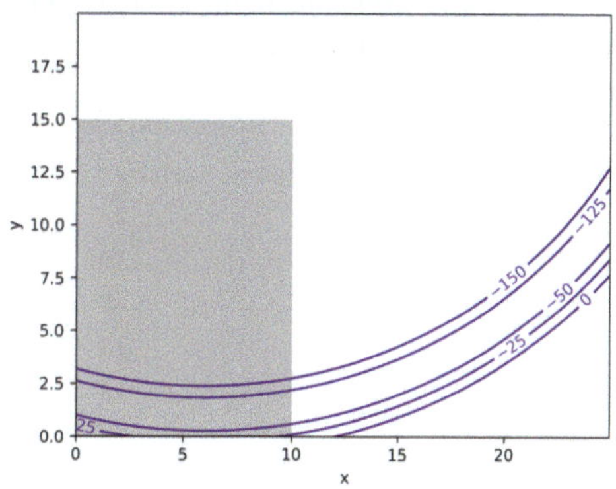

Fig. C.2 Figure for solution to Exercise 11.3

Chapter 12

12.1 The linear programming problem to be solved is

- Maximize the objective function

$$f = x_6$$

- Subject to the constraints

$$2x_1 + x_2 - 2x_3 + 3x_4 + x_5 - x_6 \geq 0$$

$$x_1 + 2x_2 + 6x_3 - 3x_4 + 2x_5 - x_6 \geq 0$$

$$4x_1 + x_2 + 3x_3 + 5x_4 + 2x_5 - x_6 \geq 0$$

$$2x_1 + x_2 - 2x_3 + x_4 + x_5 - x_6 \geq 0$$

$$x_1 + x_2 + x_3 + x_4 + x_5 = 1$$

$$x_i \geq 0 \text{ for } i = 1, \ldots, 5.$$

An optimal solution is given by $x_6 = 1.\overline{55}$ which is achieved when $x_1 = 0.\overline{88}, x_2 = 0$, $x_3 = 0.\overline{11}$, $x_4 = 0$ and $x_5 = 0$.

12.2 The idea is to model the problem using a 3×3 matrix game using the following three strategies:

- Option 1—develop both products simultaneously.
- Option 2—develop product A first.
- Option 3—develop product B first.

Let p_{ij} denote the average of the per cent increase to manufacture 1 from product A and the % increase to manufacture 2 from product B when manufacturer 1 uses strategy i and manufacturer 2 uses strategy j. The payoff matrix is then described in Table C.1.

The rows correspond to the strategy of manufacturer 1 and the columns correspond to the strategy of manufacturer 2.

Table C.1 Table for solution to Exercise 12.1

	1	2	3	row min
1	8	10	10	8
2	4	-4	13	- 4
3	4	13	-4	- 4
col max	8	13	13	8

Table C.2 Table for solution to Exercise 12.3

Alternative	Sell 12 cases	Sell 13 cases	Sell 14 cases	Sell 15 cases
Buy 12 cases	132	132	132	132
Buy 13 cases	125	143	143	143
Buy 14 cases	118	136	154	154
Buy 15 cases	111	129	147	165
Prior Probability	0.1	0.3	0.4	0.2

Table C.3 Table for solution to Exercise 12.3

Alternative	Buy 12 cases	Buy 13 cases	Buy 14 cases	Buy 15 cases
Exp. Profit	132	141.2	145	141.6

The minimum of the column maxima and the maximum of the row minima is 8. So, both manufactures should use Strategy 1—thus manufacturer 1 will increase its share by 8%.

12.3 (i) Aside from the prior probability row, the entries in the payoff table, Table C.2, indicate the profit for a given alternative and the resulting situation

(ii) We choose the alternative which has minimum payoff largest. So, the decision is to purchase 12 cases.

(iii) The manager will be able to sell 14 cases with highest probability and the maximum possible profit from selling 14 cases is earned when buying 14 cases. So, according to the maximum likelihood criterion the decision is to purchase 14 cases.

(iv) The expected profits are summarized as in Table C.3.

Therefore according to Bayes' decision rule, the decision should be to purchase 14 cases.

Chapter 13

13.4 Apply Theorem 13.13 to decide which degree sequence is graphic.

13.5 Consult the discussion following the statement of Theorem 13.7.

References

1. M.L. Balinski, A.W. Tucker, Duality theory of linear programs: a constructive approach with applications. SIAM Rev. **11**, 347–377 (1969)
2. R.G. Bland, New finite pivoting rules for the simplex method. Math. Oper. Res. **2**(2), 103–107 (1977)
3. J. Brinkhuis, *Convex Analysis for Optimization: A Unified Approach* (Spring, Cham, 2020)
4. R. Bronson, G. Naadimuthu, *Schaum's Outline of Operations Research* (McGraw-Hill, New York, 1997)
5. M.L. Bynum, G.A. Hackebeil, W.E. Hart, C.D. Laird, B.L. Nicholson, J.D. Siirola, J.-P. Watson, D.L. Woodruff, *Pyomo-Optimization Modeling in Python* (Springer, Cham, 2021)
6. V. Chvátal, *Linear Programming* (W. H. Freeman and Company, New York, 1983)
7. G.B. Dantzig, A. Orden, P. Wolfe, The generalized simplex method for minimizing a linear form under linear inequality restraints. Pacific J. Math. **5**, 183–195 (1955)
8. M. Ehrgott, *Multicriteria Optimization* (Springer, Berlin, 2005)
9. H.A. Eiselt, C.-L. Sandblom, *Linear Programming and its Applications* (Springer, Berlin, 2007)
10. D. Gale, Linear programming and the simplex method. Notices Am. Math. Soc. **54**(3), 364–369 (2007)
11. J.M. Harris, J.L. Hirst, M.J. Mossinghoff, *Combinatorics and Graph Theory* (Springer, New York, 2008)
12. F.S. Hillier, G.J. Lieberman, *Introduction to Operations Research* (McGraw-Hill Education, New York, 2022)
13. D.G. Luenberger, Y. Ye, *Linear and Nonlinear Programming* (Springer, Cham, 2021)
14. R.T. Rockafellar, *Convex Analysis* (Princeton University Press, Princeton, 1970)
15. B.E. Sagan, *Combinatorics: The Art of Counting* (American Mathematical Society, Providence, 2020)
16. J.K. Strayer, *Linear Programming and its Applications* (Springer, New York, 1989)
17. W.L. Winston, *Operations Research: Applications and Algorithms* (Duxbury Press, Boston, MA, 1987)
18. G. Ziegler, *Lectures on Polytopes* (Springer, New York, 1995)

N. Grieve, *An Introduction to Mathematical Programming and Network Science*, Springer Undergraduate Texts in Mathematics and Technology,
https://doi.org/10.1007/978-3-032-13330-4

Index

N. Grieve, *An Introduction to Mathematical Programming and Network Science*,
Springer Undergraduate Texts in Mathematics and Technology,
https://doi.org/10.1007/978-3-032-13330-4

D

E

F

G

H

I

K

L

M

N

MIX
Papier aus verantwortungsvollen Quellen
Paper from responsible sources
FSC® C105338

FSC
www.fsc.org

If you have any concerns about our products,
you can contact us on
ProductSafety@springernature.com

In case Publisher is established outside the EU,
the EU authorized representative is:
Springer Nature Customer Service Center GmbH
Europaplatz 3, 69115 Heidelberg, Germany

Printed by Libri Plureos GmbH
in Hamburg, Germany